*Human Communication
on the Internet*

Human Communication on the Internet

Leonard J. Shedletsky
University of Southern Maine

Joan E. Aitken
University of Missouri—Kansas City

Boston • New York • San Francisco
Mexico City • Montreal • Toronto • London • Madrid • Munich • Paris
Hong Kong • Singapore • Tokyo • Cape Town • Sydney

Executive Editor: *Karon Bowers*
Series Editorial Assistant: *Jennifer Trebby*
Marketing Manager: *Mandee Eckersley*
Senior Editorial-Production Administrator: *Beth Houston*
Editorial-Production Service: *Walsh & Associates, Inc.*
Compostion and Prepress Buyer: *Linda Cox*
Manufacturing Buyer: *JoAnne Sweeney*
Cover Administrator: *Kristina Mose-Libon*
Electronic Compostion: *Modern Graphics*

For related titles and support materials, visit our online catalog at www.ablongman.com.

Copyright © 2004 Pearson Education, Inc.

All rights reserved. No part of the material protected by this copyright notice may be reproduced or utilized in any form or by any means, electronic or mechanical, including photocopying, recording, or by any information storage and retrieval system, without written permission from the copyright owner.

To obtain permission(s) to use material from this work, please submit a written request to Allyn and Bacon, Permissions Department, 75 Arlington Street, Boston, MA 02116 or fax your request to 617-848-7320.

Between the time Website information is gathered and then published, it is not unusual for some sites to have closed. Also, the transcription of URLs can result in typographical errors. The publisher would appreciate notification where these errors occur so that they may be corrected in subsequent editions.

Library of Congress Cataloging-in-Publication Data

Shedletsky, Leonard, 1944-.
    Human communication on the Internet / Leonard J. Shedletsky, Joan E. Aitken.
       p. cm.
    Includes bibliographical references and index.
    ISBN 0-205-36031-9
       1. Internet—Social aspects. 2. Communication and technology. I. Aitken, Joan E. II. Title.

HM851.S537 2003
303.48′33—dc22

2003056332

Printed in the United States of America

10 9 8 7 6 5 4 3 2 1   09 08 07 06 05 04 03

In memory of Lenny's brother, Murray Sheldon

Contents

PART ONE • *Foundations*   1

1   *Introduction: Underpinning Ideas*   3

Focus   4

Purpose   6

Interactivity   6

End Users   7

Paradoxical Implications   7

Reflection   11

Case for Discussion   12

Internet Investigation   12

Concepts for Analysis   13

2   *Process of Human Communication on the Internet*   15

Focus   16

Developmental Perspective   17
     Where Do We Start?   18
     Cyberliteracy   21

Why Study Human Communication on the Internet?   23

Communication Media   25
     Language and Writing   25
     Visual Medium   26

Models of Communication   28

Reflection   30

Case for Discussion   30
Internet Investigation   31
Concepts for Analysis   32

3 Tensions of Communication on the Internet   34

Focus   34

Globalization   36

Cyberspace   36

Theoretical Roots   37

Humanist versus Behavioralist Approach   38
    Meaning   41

Adoption and Diffusion   41

Informing and Conversing   42

Reflection   43

Case for Discussion   44

Internet Investigation   44

Concepts for Analysis   44

PART TWO • *Functions*   47

4 Informatics   49

Focus   49

What Is Informatics?   51

Research Communication   52

Scholarly Communication   53
    Refereed Journals   54
    Critical Reading and Peer Review   54

Quality Primary Sources about Communication   55

Online Databases   55
    Full-Text Databases   57

Skill Section: Online Search   57

E-Journals   60

Journalism and Internet Scholarship   61

Communicating Scholarly Research Online   62

Online Scholarly Discussion Groups   63

Reflection   64

Case for Discussion   65

Internet Investigation   66

Concepts for Analysis   67

5 The Play of Internet Communication   70

Focus   71

The Play of the Internet   72

    Imaginative Play Like a Child   75
    Identity Role-Playing   75
    Interplay with the Self or Other   76
    Play Like Theater   76
    Games for Destruction   76
    Playful Work   77
    The Fun of Something New   77

Play of Computer Games   79

Playful Metaphors   79

Example Metaphor   81

Interplay between Internet and Communication Principles   84

    Internet Communication Is a Process   84
    Adapting to the Internet Will Increase Communication Effectiveness   84
    Communication Is Irreversible   84
    Internet Meanings Are in People   85
    No One Can Experience Totally Effective Communication on the Internet   85

Skill Section: Communication Design on the Web   86

Reflection   87

Case for Discussion   88

Internet Investigation   88

Concepts for Analysis   89

6 Polarization of People 91

Focus 92
Intensification of Polarization 94
Digital Divide or Technical Together? 95
Cognitive Dissonance 97
Video Games 98
Privacy 99
Hoaxes, Rumors, and Myths 100
Internet Addiction 101
Hostile Metaphors 102
Flaming 105
    What Stimulates Flaming? 105
Personal Fear 106
Hate Speech 107
Fear of Terrorism 109
Facing Fears 110
Reflection 111
Case for Discussion 112
Internet Investigation 112
Concepts for Analysis 113

PART THREE • *Communication Modes* 115

7 Intrapersonal Communication as Cognitive Collaboration 117

Focus 118
Intrapersonal Communication and the Internet 119
    Inner/Outer Speech 122
Resolving the Inner/Outer Dichotomy 124
The Self and the Internet 127

Attributing Human Characteristics to Computers 130

Interactive Communication 131

Role-Playing Identity 132

Culture 133

Age 134

Gender 136

Reflection 137

Case for Discussion 138

Internet Investigation 138

Concepts for Analysis 139

8 Interpersonal Communication on the Internet 141

Focus 141

Relationship between Interpersonal and Internet Communication 143
- Meeting People and Creating Relationships 146
- Enhancing Relationships with Family and Friends 146
- Maintaining Long-Distance Relationships 146

Skill Section: Speed, Reach, Anonymity, Regard, and Interactivity 149

Influence on Family 152
- The Internet as a Source of Communication Content 153
- Using the Internet to Increase Human Interplay 153
- Reinforcing a Child's Self-Esteem 154
- Enabling New Opportunities for Interpersonal Relationships 154

Online Relationships 155

Reflection 158

Case for Discussion 161

Internet Investigation 161

Concepts for Analysis 163

9 Groups 164

Focus 164

Mediator of Human Communication 166

Connecting People Who Have Needed Information or Resources   166
Empowering the Self through Support Information and Self-Expression   167
Improving the Quality of Life   168

Storytelling in Groups   **168**

Developing Community through Groups   **171**

Characteristics of Online Group Discussion   **172**
Unique Language   173
Ignoring   174
Lurking   174
Conversation Lulls   174
Skill Section: High-Volume E-mail   175
Internet Anonymity   175
Virtual Reality   176
Lack of Nonverbals   176

Skill Section: Considerations for Moderating Online Discussion   **177**
Consider Creating a Pragmatic System in Advance   177
Consider Encouraging Brief E-Mails   177
Consider Avoiding Private, Direct E-Mails   178
Consider Allowing Confidential Talk   178

Health Care Empowerment through Groups   **178**
Patient-Doctor Communication   179
Patient Storytelling Groups   179
Accuracy of Information   179

Reflection   **180**

Case for Discussion   **181**

Internet Investigation   **181**

Concepts for Analysis   **182**

PART FOUR • *Contexts outside the Home*   185

10   *Workplace Contexts*   187

Focus   **187**

Perspectives of Understanding Organizational Communication   **188**

Revising Traditional Theories about the Workplace   **189**
Bureaucracy   189
Scientific Management   189

Administration   190
Classical Management   190
Human Relations   190
Theory X and Theory Y   190
Participative Decision Making   191
Symbiotic Relationship between People and Organizations   191
Contingency Theories   191

Business Communication on the Internet   **191**

Systems   **192**

Communication Patterns   **193**

Roles   **193**
Tensions   195
Paradoxes Abound Regarding the Internet and Organizations   195
Decision Making   195
Internet Play May Interfere with Business   196

Culture and Climate   **197**

E-Business   **199**

Online Business Interaction   **200**
Consumer Communication   200
Interpersonal Communication   201
Collaborative Communication   201
Groupthink   202
Communication Overload   202
Risky Communication   203

Skill Section: Manager Communication   **203**

Reflection   **204**

Case for Discussion   **204**

Internet Investigation   **205**

Concepts for Analysis   **206**

11 *Educational Contexts*   209

Focus   **209**

Intensified Learning   **211**

Allow Anyone to Learn Anything, Anytime, Anywhere   **212**

Intensification of Effort   **215**

Educational Policy and Internet Communication 217

Academic Collaboration 217

Reflection 219

Case for Discussion 221

Internet Investigation 221

Concepts for Analysis 222

PART FIVE • *Implications* 223

12 *Consequences and Conclusions* 225

Focus 226

Convergence 228

Communication Perspective 229

In the Works 234

Reflections 236

Skill Section: Internet Behavior 240
- Confirm Important Communications 241
- Engage Your Sense of Humor 241
- Ask Questions 241
- Use Netiquette 241

Reflection 244

Case for Discussion 245

Internet Investigation 245

Concepts for Analysis 246

References 247

Index 263

Preface

We wrote this book to better understand how communication on the Internet works. We had a personal reason to understand how it works. We were early adopters of the Internet. Since the early 1980s we have been using the Internet and integrating it into our classes. We have been exploring the Internet as a medium of communication. Before we ever met face-to-face, in the mid 1990s, we collaborated in co-editing a book via the Internet. At the same time, we found ourselves trying to explain how communication on the Internet works. We began to observe how people learn to use the Internet, what they seek from it, what they dislike about it, how they use it. We had written about the cognitive or intrapersonal processes of communication on the Internet; we had written about interpersonal processes of communication on the Internet; we had written about the paradoxes we observed in Internet communication. We wanted to follow those ideas to see where they would go.

Even before the days of point and click, when we had to type commands to get computers to reach out over the network of computers, we began to tell our students to think of the computer as a crayon and a piece of paper—a way to express yourself. The point of the metaphor was to emphasize expression of ideas, communication, simplicity of technology, child's play. People liked it. There was a time when we could calm students (sometimes K–12 teachers) who suffered with computer anxiety by reminding them that we ought to think of the computer and the Internet as a communication device, much like the telephone or a typewriter or a crayon. Today, it is less likely that we would need to say any of that. We find that people have come to think of the Internet as a communication instrument. This book is simply a close look at ways in which communication works on the Internet.

Early in the book, we provide the foundation for exploring the ways people communicate on the Internet. In essence, we are concerned with the same fundamental processes here as in the prototype of two people talking face-to-face. We want to know: What is going on? How does it happen? What factors influence the process of communication? Our desire is to prompt your thinking and investigation into how the Internet influences humans to function, think, communicate, learn, change, and evolve.

In Part One, Foundations, we introduce communication as a process, and we examine how the process might work within the medium of the Internet. We explore the paradoxical nature of communication on the Internet.

Chapter 1—Introduction—gives you the underpinning of ideas to guide the book. We discuss our purpose in writing the book and our desire for your interactive learning. Our focus is on the process of communication, not technology. We discuss the perspective of end users and how we can benefit from learning about communication on the Internet. We have a theoretical point of view in this book: Communication on the Internet is paradoxical.

Chapter 2—The Process of Human Communication on the Internet—gives you an overview of how the communication process works in computer-mediated communication. We discuss cyberliteracy and the importance of studying the Internet. We discuss the Internet as a communication medium, and the characteristics of language and writing, visual orientation, and intensification. We discuss a model of communication as a way to understand this abstract process.

Chapter 3—Tensions of Communication on the Internet—gives an elaboration of what we believe is the paradoxical nature of communication on the Internet. We discuss the paradoxes of globalization as an example of the different forces operating on the Internet. In discussing cyberspace, we apply theories of interpersonal, dialectical tensions to online communication. In fact, we believe online communication is more like interpersonal communication than it is like any other mode of communication. We discuss tensions related to meaning-making, adoption, and diffusion. As we lead into the next section, we discuss the distinction between conversing and informing and relate that distinction to online communication.

In Part Two, Functions, we discuss basic functions of communication via the Internet. We see three major functions: conversing, informing, and playing. As part of the paradoxical nature, however, the Internet actually can polarize people.

Chapter 4—Informatics—gives you ideas about how the Internet is used as a communication vehicle for information. We discuss research and scholarly communication, including the value of critical reading and peer-reviewed journals. You will learn about the importance of using quality, primary sources. And you'll learn how to find those sources through databases and search engines.

Chapter 5—The Play of Internet Communication—gives insight into the ways people play online. We discuss computer games, metaphors, and other types of interplay between people and communication principles.

Chapter 6—Polarization of People—gives you a perspective of much of the dark side of communication. The Internet can cause a digital divide, dissonance, and violence. People can use the Internet to violate your privacy, spread rumors and hoaxes, steal, and more. We talk about the nature of flaming online and suggest reasons why we think flaming occurs. We also explore the use of the Internet by hatemongers and terrorists.

In Part Three, Communication Modes, we explore communication on the Internet through the perspectives of three communication modes: intrapersonal, interpersonal, and group. We stopped short at public speaking and mass communication because we believe communication on the Internet to be primarily a conversing or interpersonal mode. We do, however, believe that public and mass communication frequently occur on the Internet, although with the Internet, any type of communication can become public!

Computer-mediated public communication includes teleconferencing, online meetings, public archives of e-groups, and web sites. But we find that the Internet blurs modes of communication. In fact, the various modes converge into a new mode of communication: Internet-mediated communication.

Chapter 7—Intrapersonal Communication as Cognitive Collaboration—gives you an explanation of how people come together mentally through their online conversing.

Chapter 8—Interpersonal Communication on the Internet—gives insight into how people build and maintain relationships online.

Chapter 9—Groups—gives perspective on how communities emerge online.

In Part Four, Contexts outside the Home, we examine the contexts of Internet communication or where the communication takes place outside the home. Specifically, we discuss work and school contexts.

Chapter 10—Workplace Contexts—gives you ideas about systems, roles, tensions, and decision making influenced by the Internet. We discuss influential factors of organizational climate. We also explore some characteristics of e-commerce.

Chapter 11—Educational Contexts—gives perspectives about Internet communication in schools, particularly higher education. The Internet intensifies learning. It allows anyone to learn anything, anytime, anywhere. The Internet intensifies effort and collaboration.

Chapter 12—Consequences and Conclusions (Part Five, Implications)—is unique in that we attempt to point you toward the implications of communication on the Internet. We summarize ideas we have discussed throughout the book and ask you to contemplate future directions.

In sum, this book is our attempt to describe communication on the Internet. We rely upon intrapersonal communication theory, interpersonal communication theory, and Internet communication theory, all the while attempting to paint a picture that is coherent and persuasive, one we can recognize from our experiences. This is a picture that is subject to your judgment. We invite you to join in. Read, investigate, and interact.

Acknowledgments

Any book is the combined effort of many people. We appreciate the enthusiastic support of our editor, Karon Bowers, and her assistant, Jennifer Trebby. We thank the people who contributed to the research in this book—subjects who completed surveys and agreed to interviews, our colleagues who discussed their ideas with us, the many people we met online who were willing to let us discuss their virtual world, and our friends who enjoyed telling us how they communicate online.

Joan thanks the staff at the Miller Nichols Library of the University of Missouri-Kansas City. We received WebCT help from Sheryl E. Coleman and Jesse L. Whitlock (UMKC-TLTC staff). Joan's favorite librarian, Frank E. Dalrymple, added explanations about online databases. Lenny relied heavily on the University of Southern Maine's Ron Levere, Director of Instructional Technology and Media Services, for everything from moral support to software support. Also, thanks to Jane Danielson, Instructional Designer, for Blackboard help and Scott Kimball, Instructional Media Support Specialist, for his graphics work on the intrapersonal communication model.

We are grateful to the Allyn and Bacon reviewers for their time and input: Dudley Cahn, State University of New York at New Paltz; Stephen D. Cooper, Marshall University; Michael Dreher, Bethel College; John R. Foster, Northwestern State University of Louisiana; Laura Gurak, University of Minnesota; Lynne Kelly, University of Hartford; Elizabeth Leister, LaSalle University; John Malala, University of Central Florida; Edward McGlone, Emporia State University; William S. Stone, Jr., Northeast Mississippi Community College; Fay Sudweeks, Murdoch University; and Joseph B. Walther, Cornell University.

A special note of gratitude goes to the family of Joan's colleague and friend Clinton W. Ferry, the manager of Computer Resources for the College of Arts and Sciences at the University of Missouri-Kansas City. For sixteen years, Clint was Joan's main computer guru at the office. For Clint, no computer question was ever too mundane. For Clint, any computing problem could be solved. With little help and too few resources, Clint was able to exemplify the best in computing support. Joan is deeply saddened by Clint's unexpected passing during the week before the completion of this manuscript.

Of course, the patience and help of our families is essential to our success. We are fortunate to have loving life partners who are well-trained and practiced in computing. Lenny's wife, Cathie Whittenburg, provided computer help and often pulled Lenny from the darkness of error messages. Lenny's brother-in-law, Jack Whittenburg, built Lenny's computer and fixed all those technical problems that seemed insurmountable. Joan's husband, Rodger D. Palmer, supplied essential computer assistance. We also found the computer knowledge and perceptions of our children helpful in shaping this book. Thus, we thank Lenny's children, Jo Temah and Noah, and Joan's children, Copper and Wade Aitken-Palmer.

Finally, and most importantly, we thank our students. Our work is always motivated by the students we teach. They remind us what it means to teach and to learn. They reward us for being clear. They allow us to join them in learning, no matter how old we get.

Part I

Foundations

Early in this book, we provide the foundation for exploring the ways people communicate on the Internet. Our desire is to prompt your thinking and investigation into how the Internet influences humans to function, think, communicate, learn, change, and evolve.

Chapter 1—Introduction—Gives you the underpinning of ideas to guide the book. We discuss our purpose in writing the book and our desire for your interactive learning. Our focus is on the process of communication, not technology. We discuss the perspective of end users and how we can benefit from learning about communication on the Internet. We have a theoretical point of view in this book: Communication on the Internet is paradoxical.

Chapter 2—The Process of Human Communication on the Internet—Gives you an overview of how the communication process works in computer-mediated communication. We discuss cyberliteracy and the importance of studying the Internet. We discuss the Internet as a communication medium, and the characteristics of language and writing, visual orientation, and intensification. We discuss a model of communication as a way to understand this abstract process.

Chapter 3—Tensions of Internet Communication—Gives an elaboration of what we believe is the paradoxical nature of communication on the Internet. We discuss the paradoxes of globalization as an example of the different forces operating on the Internet. In discussing cyberspace, we apply theories of interpersonal, dialectical tensions to online communication. In fact, we believe online communication is more like interpersonal communication than it is like any other mode of communication. We discuss tensions related to meaning-making, adoption, and diffusion. As we lead into the next section, we discuss the distinction between conversing versus informing, and relate that distinction to online communication.

1

Introduction: Underpinning Ideas

> *"Dang! I hate to sound so uncritical, but this manuscript knocks me out."*
> —(Anonymous Reviewer)

Granted, we started with this quotation because we are still basking in the reviewer's glowing reaction to our work. We also started with this quote, however, to give you insight into our point-of-view in studying human communication on the Internet.

```
Date: Mon, 11 Nov 2002 15:54:16-0500
To: "J. Aitken" <aitkenje@hotmail.com>
From: Lenny Shedletsky <lenny@maine.edu>
Subject: did you notice?
- - - -
the reviewer used the term, "DANG." is that cosmic or
what?
```

We both noticed that this reviewer—like both of us—uses "Dang!" We found the comment playful and intense, as "Dang!" happens to be a favorite reaction we use with each other as we have studied communication together over the years. We use the word "Dang!" to show "Ah Hah" moments, to reward one another with a cute-sounding

3

superlative, to playfully remind one another about an earlier period in our lives when that term was part of the vernacular of the day. The word reminds Lenny of the years he spent in the Midwest. Does our reaction to the word, "dang," reflect something about our commonality? Our language use? Or perception? Our age? As communication scholars, we analyze language, meaning, and context. Thus, you can begin to understand our perspective of analyzing meaning-making of any communication message within the medium of the Internet. By the way, we suggest you refer to us as "Lenny and Joan" because few people can pronounce either of our last names correctly and there is an expected equity and familiarity between people on the Internet.

Focus

```
TO: Students

FROM: paradoxdoc@hotmail.com

RE: Overview: Introduction
```

In the introduction, Lenny and Joan give you their perspectives for exploring the paradoxes of human communication on the Internet. Language and meaning-making are central factors in their understanding of the Internet. They discuss the interconnected blur of intrapersonal, interpersonal, and mass communication interaction through the Internet. The variables of play, paradox, and heightened communication are some of the concepts they value in their analysis of communication on the Internet.

Lenny and Joan invite you to join them in thinking about the human mind and body and how the mind and body work in communication. They are trying to understand how the Internet changes us, our communication, our society, and the world.

Their book provides ideas to stimulate discussion and probing, so that you can ponder, discuss, and react to events you encounter on the Internet as part of your investigation. Lenny and Joan want you to consider the Internet from the standpoint of an end user, although you will find value in other perspectives.

They discuss the paradoxical nature of human communication on the Internet. Does the Internet create a global village or retribalization, for example? Does the

> Internet make you feel better connected to others or more isolated and depressed? Yes! Lenny and Joan believe that paradox is in the nature of human communication generally and communication with technology in particular.
>
> In their introduction, they discuss two types of convergence. First, there is the convergence of technology. Second, they found it increasingly difficult to distinguish between the intra, inter, group, and mass communication modes.
>
> Finally, because of the interactive nature of the Internet, Lenny and Joan encourage you to use the book, website, and course environment in an interactive learning approach. They hope you will read the book, explore the Internet, do your own review of the research about the Internet, conduct original research, and voice your ideas to your class and to the book's web site as well.

In this chapter, we explore our perspectives on the study of communication on the Internet. We fully recognize that students sometimes avoid reading the introduction to a book. We hope you will read this chapter as a way of underpinning the ideas of the book. At the end of this chapter, you should be able to answer the following questions.

1. What is the purpose of this book?
2. What are interactive approaches to learning in this course?
3. Why are there paradoxes about human communication on the Internet?
4. What are some of the paradoxes that exist about human communication on the Internet?
5. What strategies will you use to think critically while reading this text?

This book is an exploration of the paradoxes of human communication on the Internet. We want to stimulate your thinking about the qualitative aspects of communication on the Internet. We want to speculate about the nature of human communication on the Internet and the possible influence of the Internet on our communication lives, taking into account the characteristics of both human communication and the Internet. This book is about communication, not Internet technologies themselves. The text is not a "how to" book, although there are places where we do suggest how to do things (see Skill Sections). Nor is this book a summary of all the empirical findings we could gather together into one place—e.g., How many people do this or that? How often? For how long? Certainly, we will make reference to statistics and social science findings, but our aim is broader. Our desire is to prompt your thinking and investigation into how the Internet influences humans to function, think, communicate, learn, change, and evolve, and how humans influence the Internet.

Purpose

Our goal is for you to join us in thinking about the human mind and body and how the mind and body work in communication on the Internet. Our goal is not for you to memorize what we write here. Ask yourself: How does the Internet change us? We welcome your input. In fact, that is why we introduce you to academic research on the Internet in Chapter 4, Informatics, offer an online appendix with databases, and encourage you to use an interactive course environment with the text. We think it is important that you take note of your ideas and examples from your experiences on the Internet. Supporting materials include a glossary of terms available with this book, and course support through course environments. We will assume that as you read this book, you will explore the Internet, at times as a communicator and at times as an observer of the communication event.

Our intent is to provide ideas to stimulate discussion and probing by you. We prepared this book as if we were initiating discussion in our own classes. This book is limited to a set of topics because our plan is not to cover a body of information, but to engage your imagination. Key to our approach is the online exploration you will do; discussion you will engage in on your course discussion board; and research you will carry out, both hands-on and literature review. Therefore, we strongly encourage you to take special note of the E-Group Discussion and Internet Investigation sections in each chapter.

You will notice that this text has definitions at the end of each chapter and an extensive glossary of terms and database information available through the course environments—*Course Compass* and *Blackboard*. Faculty can select their preferred course environment, then adapt the materials for their particular course. Individual case studies and skill sections also have been included to help you understand the impact of the Internet on human communication. Throughout this book, you will find anecdotes about people we interviewed or surveyed, although we changed the names of the people involved. Sometimes such an example is more extreme than representative, but each suggests the Internet's potential influence on human communication.

Interactivity

We invite you to observe the communication events you encounter on the Internet much as we would invite you to look at your interpersonal encounters if you were enrolled in a course on interpersonal communication; or your small group experiences if you were enrolled in a course on small group communication; or your own intrapersonal communication if you were enrolled in a course on the communication that happens within your self. Our aim is to involve you in an investigation rather than to report exhaustively on the research literature. The book attempts to touch upon interesting and important areas to think about as we try to give you some tools to begin this adventure.

The approach we take in this book is to consider the oppositions, dualities, dichotomies, tensions, paradoxes, or contradictions that appear to operate as we communicate over the Internet. We point out paradoxes to stimulate thought, imagination, and exploration. As Vitanza (1999) has pointed out, the first irony or paradox here is that we are discussing the electronic age in a traditional book. He called this period in our communication history proto-electronic (p. 231). What would you call this point in communication history?

To find paradoxes of human communication on networked computers, we need not look any farther than the soul-searching of a software engineer who writes:

> Still, this time I'm working on a "good" project, I tell myself. We are helping people, say the programmers over and over, nearly in disbelief at their good fortune. Three programmers, the network guy, me—fifty-eight years of collective technical experience—and the idea of helping people with a computer is a first for any of us. Yet I am continually anxious. How do we protect this database full of the names of people with AIDS? Is a million-dollar computer system the best use of continually shrinking funds? It was easier when I didn't have to think about the real-world effect of my work. It was easier—and I got paid more—when I was writing an "abstracted interface to any arbitrary input device." When I was designing a "user interface paradigm," defining a "test-bed methodology." I could disappear into weird passions of logic. I could stay in a world peopled entirely by programmers, other weird logic-dreamers like myself, all caught up in our own inner electricities. It was easier and more valued. In my profession, software engineering, there is something almost shameful in this helpful, social-services system we're building. The whole project smacks of "end users"—those contemptible, oblivious people who just want to use the stuff we write and don't care how we did it. (Ullman, 1997, pp. 8–9)

End Users

"End users." That's us, at least the two of us writing this book. Our expertise is in analyzing how end users communicate. We are trying to understand the electronic world in which we live, the world in which we act and react, the world of communicating over the Internet. We think that one way to make sense out of this complicated and quickly evolving world is to recognize the many paradoxes or seeming contradictions in that world. In Ellen Ullman's description above, of a software engineer's thoughts on her work, one can sense a tension between the pull to stay within the framework of the logic of programming and computers and the framework of human communication, including human, moral, medical, and social needs. The software engineer finds herself pulled in opposing directions simultaneously. In trying to help, she may cause harm. In doing work to be used by people, end users, she finds contempt for the end users.

Paradoxical Implications

James Carey, a distinguished communication theorist, was asked whether communication technology is expanding human experience, enhancing our quality of life, and facilitating democracy, or just the opposite (Wilhelm, 2000); reinforcing barriers between people and diminishing social interaction (Arakaki Game, 1998). Carey began to respond by rejecting talk either hailing contemporary technology or denouncing technology. Instead, he observed, "The consequences of technology are always profoundly contradictory; contradiction is of the essence of technology, not just some accidental byproduct of the historical process" (Arakaki Game, 1998, p. 127). As Steven Jones (1997) explains, "That the adoption of technology can have an effect opposite to the one intended should not be surprising, for we have become accustomed (perhaps from the very first time we must deal with the consequences of a thunderstorm that has cut power to our area and left us without refrigeration, lights, air conditioning, television, etc.) to the "trade-offs" that occur as we develop and implement technology" (p. 8).

Referring to the now common term, "global village," associated with Marshal McLuhan's probes into the electronic age (we had no Internet in 1964), Herring (2001), points out that the "global village" is a term that itself is oxymoronic, that is, seemingly contradictory. "Global" brings to mind the whole world. "Village" stimulates images of a small, local community of people and buildings. But, if you accept for a moment the idea of a global village—entailing messages and forms of communication widely distributed—then you find some people who see this communication as a means to reducing cultural isolation and distance. Other people see the Internet as a new and powerful way for the United States to bring about *cultural homogenization,* which is domination of U.S. culture over other cultures (Ess, 2001; Mowlana, 1995). Still others see the Internet as a way of retribalizing the world because all the individual user groups are comparable to tribes (Van & Mandel der Leun, 1996). We hear of other paradoxes. Consider, for example, the employer's concern that where the Internet was supposed to increase productivity, the Internet has instead enabled workers to socialize at work or check their investments online and thus has actually lowered productivity (Rice & Love, 1987; Schmitz & Fulk, 1991). We will raise other paradoxes throughout the text.

An interesting line of empirical research has been spawned by exploring what some researchers have referred to as the Internet paradox hypothesis (Kraut, Patterson, Lundmark, Kiesler, Mukopadhyay, & Scherlis, 1998; LaRose, Eastin, & Gregg, 2001). Empirical research is a systematic investigation of the observable. These researchers recognized that a dominant use of the Internet is for interpersonal communication (Kraut, Mukhopadhyay, Szczypula, Kiesler, & Schleris, 2000), and that we cannot generalize whether the Internet will promote or reduce interpersonal connection and promote or reduce psychological well being. Some researchers think the Internet will increase depression and loneliness and others think the Internet will reduce depression and loneliness. Some researchers think the Internet will improve social ties and others think the Internet will worsen social ties. Researchers recognize that the technology could be viewed as doing either—helping or hurting—or both simultaneously. While Kraut, Patterson, Lundmark, Kiesler, Mukopadhyay, and Schleris (1998) found evidence to support the adverse effects of the Internet on social involvement and psychological well-being (loneliness and depression), other research and theorizing has called that finding into question. In fact, a number of other studies support the positive side of the paradox, that the Internet enhances our communication lives (Hamman, 1999; Riphagen & Kanfer, 1997; Walther, 1996; Parks & Floyd, 1996; LaRose, Eastin, & Gregg, 2001). Lest you feel discouraged by these opposing research findings, you should know that this paradoxical nature is not an uncommon state of affairs in the world of research. What paradox signals—short of instances where research is simply invalid—is that:

1. We do not yet understand the Internet, or
2. We cannot generalize the highly personalized effects of the Internet to everyone, or
3. The Internet has paradoxical effects on each individual.

In the few years since the Internet began, the theories, hypotheses, and research have not been able to discriminate the "forces" that are operating. In addition, the rapidly changing and evolving nature of the Internet can make yesterday's conclusions questionable today. What we are suggesting is an additional take on paradoxes, namely, that paradox is in the nature of human communication generally and communication with technology in particular.

In other words, in this book, we explore the idea that communication on the Internet is characterized by its paradoxical[1] or contradictory consequences. The paradoxes do not reflect a lack of understanding, but an actual increase in understanding and acceptance of the reality of the virtual world. The strategies, values, and effects of communication on the Internet contain inconsistencies, opposing forces, and illogic. We propose that the paradoxical nature of communication on the Internet and the increased awareness of the intrapersonal processes of meaning-making intensify the experience of communication.

Researchers who take a close look at just how people talk to one another—following the methods of discourse analysis or conversation analysis, or what some scholars have referred to as practice theory—examine the details of talk (Bakhtin, 1981; Basso, 1992; Bauman, 1975, 1992; Gumpertz, 1982; Miller & Hoogstra, 1992; Ortner, 1984). These researchers observe word choice, turn-taking, implied meaning, social actions enacted, and much more. Because ordinary talk both constructs a linguistic and cultural context and takes on meaning within various levels of context, online discussion is a particularly appropriate arena for such analysis. Stripped of everyday clues to context, we must work all the harder to create a way to accurately interpret meaning, a process made more difficult because we cannot observe online context as a way to infer meaning and recognize intentions (Baym, 2000).

Theorists have long suspected that the written form of language has had an influence on how we think. Baron (2000) reminds us that Walter Ong (1982) held such a view of writing: "In Ong's words, 'by distancing thought, alienating it, from its original habitat in sounded words [that is, speech] writing raises consciousness'" (pp. 301–302). By consciousness we mean awareness through perceptual understanding. But as you will read more about later on in this book, much of written communication on the Internet is a mix of writing and speech—a new form, as we find in e-mail. In addition, much of communication on the Internet in this new hybrid combination of the oral and the literate is also a simulation of conversation. Not quite the same as face-to-face conversation, but similar. So, at the level of communication form, we encounter a convergence in interactive modes of the Internet.

We observe two types of Internet convergence. First, there is the convergence of technology. The media are becoming one medium. When we look at our computer screen, we have a stereo, television, telephone, Internet service, and other media technology all converging together. Second, we have an increasingly difficult time distinguishing between our intra, inter, group, and mass communication modes. On the Internet, there is a convergence and synergy of *intrapersonal communication, interpersonal communication,* and *mass communication,* which intensifies the experience of the communicator. The heightened intensity of the communication event may be likened to the idea of arousal stirred by mass media—the idea that attending to a particular message may increase a general emotional arousal, as described in excitation transfer theory (Zillmann, 1971).

Not surprisingly, we find paradoxes in the communication event. At the same time that the Internet brings us together in an experience of one another, in an intimate exchange of meaning, the Internet erects barriers between us. At the same time as the Internet helps to create community, communication on the Internet facilitates the turning away from the geographically defined local and embodied community. As the Internet helps to build community online, communication on the Internet emphasizes the individual in social relations (Fernback, 1997). While the Internet extends our reach in discussion with others for work and education, communication on the Internet intrudes upon our time away from work and school. The Internet comes into our private time and space, while changing the notion of time and space. The Internet draws us inside to the workings of our minds and outside to the mass produced and distributed messages of mass communicators. The Internet offers privacy and, simultaneously, a window to make our private lives public. Communication technology itself has been simultaneously celebrated and denounced (Arakaki Game, 1998; Kraut, Kiesler, Boneva, Cummings, Helgeson, & Crawford, 2002; Shedletsky & Aitken, 2001).

Privacy. By way of example, consider the complicated issue of privacy on the Internet. When using the Internet, we gain freedom to access information, but we give up our individual privacy because of computer tracking information. According to Brail, we enjoy free speech online, but we also run the risk of harassment (Brail, 1996, cited in Peterson, 1994). The Internet itself is neither good nor bad, but human communication on the Internet can present the best of humanity and the worst.

> Along with its many benefits, the march of technology makes an encompassing surveillance network seem almost inevitable. We owe much of the privacy we have enjoyed in the past to a combination of immature technology and insufficient manpower to monitor us. But these protective inefficiencies are giving way to efficient technologies of data processing and digital surveillance that threaten to eliminate our privacy. Already we are tracked by our credit-card transactions, our travel through automated toll booths, our cell phone calls. Each year brings more sensitive and widespread sensing devices, including cameras, microphones, and, potentially, biological sensors, all of which are being connected through increasingly efficient networks to increasingly more powerful data-processing and storage devices. Cameras are proliferating: at toll plazas, on public streets, and in public parks. We welcome them as crime fighters even as they eliminate our ability to move through the world untracked. Face and voice recognition software may soon permit image data from surveillance cameras to be cross-referenced to databased profiles of each person observed. For a glimpse of the future, enter your street address at globexplorer.com. You will see a satellite picture nearly good enough to show a car parked in your driveway, or in mine. Better resolution is coming soon. We are moving toward a transparent society in which our actions and transactions are followed, our lives tracked and documented, by folks we neither know nor trust—each of us a star in our own Truman Show. (Nesson, 2001, para. 3)

Have we lost freedoms through a loss of privacy? Are we at greater risk because of the Internet? Should we feel fear because of the Internet? Certainly some individuals are more vulnerable because of the Internet. We had a student, for example, who did not want her name or photo put on the web pages for a course. The student had worked hard to escape a violent ex-husband. Her ex-husband had been imprisoned for attacking her, and the student had subsequently relocated for her safety. The student was afraid her Internet course might enable this violent man to track her down and kill her.

Contradiction. A question worth pondering is: Just what constitutes paradox and contradiction? We know that research findings are often contradictory (paradoxical) in the social sciences. Research literature often suggests that X > Y; that Y > X; that there is no difference between X and Y. Just to offer one example, this state of affairs is found in the research literature on gender differences in communication (Eagly, 1987), although you could point to many other areas to make the same argument. We generally assume in such cases that this state of affairs is due to one or more of several reasons: (a) contradictory findings are due to differences in how the research was done, just what was studied; (b) contradictory findings are due to the rudimentary state of our understanding about the phenomena under study; (c) contradictory findings are due to extraneous variables. No doubt these sorts of factors play a role in some of the contradictory findings in Internet research, but what we are especially interested in exploring is the idea that communication on the Internet produces contradictory outcomes because that is the nature of communication on the Internet, not due to error or shortcoming in our understanding.

Again, in the area of gender differences in communication, we can point to contradictions or inconsistencies of the type we seek here, where two opposing states exist simultaneously, where at one and the same time, X is different from Y, and X is the same as Y (Sayers, 1986). Experts may have difficulty knowing the difference between (a) contradiction caused by research method (methodology), (b) contradiction due to theory A versus theory B (understanding), and (c) contradiction due to the nature of communication. Therefore, we must be cautious as we explore and discuss contradiction and paradox on the Internet.

Reflection

Keep in mind as you read this book that we are teachers, and we are writing this book as if we are teaching a course on human communication on the Internet. As teachers our objective is to engage your interest and thinking about ideas. Our goal is to stimulate critical thinking about human communication and especially about human communication on the Internet. Our tendency is to view the Internet from the perspective of experts in interpersonal communication. But consider another paradox: Communication on the Internet is simultaneously interpersonal communication and mass communication. As you proceed through this text reading our perspective, ask yourself: How might an expert in mass communication view these ideas?

Because the Internet is famous for its interactive quality, and because our teaching is founded on interactivity, dialogue, or give and take, we decided to write the book as if we were initiating a dialogue with you. And you have already encountered "an e-mail" from paradoxdoc, who we think of as a helper, a narrator, or popup tutor. So, as you read this book, please do keep in mind that your learning will work best if you become an active member of this "course." We hope you will read the book, explore the Internet, do your own review of the research about the Internet, speculate, and voice your ideas on your class web site.

Case for Discussion

Dale is a management information system expert at a university. His son's soccer team asked him to create a web site for the team so they could convey information easily. Dale gladly agreed and created a multi-page team web site within the structure of the Soccer Club's site.

One day Dale received an anonymous e-mail from the soccer club's "webmaster" telling him the pages were changed to a way the webmaster liked, cutting various internal and external links, and telling Dale his page had to be consistent in style with the soccer club. Dale sent the anonymous webmaster—having no idea who the person was or what their credentials might be—a list of research-based web page design techniques, most of which Dale followed and the webmaster violated. The webmaster took control of Dale's pages, and Dale, still having no idea with whom he was corresponding, set up a soccer team web page at his personal host site.

This conflict raises questions that we will address throughout this book: Who is in control of the Internet? What is the effect of anonymous communication on the Internet? What research exists about the Internet? Is the Internet a vehicle to provide easy access to information or a way to control communication? In your course discussion group, discuss some of the problems raised by Dale's experience. Do you know how to design a web site for the Internet? Are you strictly an Internet user? What areas of expertise do you have related to the Internet? Give your analysis and perceptions of what expectations you have for learning in this course. What do you want to learn?

Internet Investigation

1. *Glossary.* Skim the review of concepts at the end of each chapter or look at the glossary in your course environment. Scan the pages and find some topics of interest to you. Note three ideas you already know and three that are new to you. Investigate the new ideas in the textbook and other sources. Post your discovery to your course online discussion group.
2. *Convergence.* In the last ten years, we have gone from few web hosting sites to thousands of web host services (Clyman, Muchmore, & Sarrel, 2003). We no longer have broadcasting, we have webcastings, videostreaming, Internet cell phones, and other technology that merge various communication media. Mass media are continually becoming more digital as they converge. Investigate convergence, the idea that we can no longer talk about just a computer, but how all media work together. Perhaps your telephone, television, computer, and Internet services are blending together. What is the difference between analog and digital media? Give examples of each. How does convergence of media affect human communication?

3. *Privacy.* Privacy on the Internet is a highly complicated and sometimes delicate concern. Investigate concerns about privacy on the Internet (e.g., downloading music, civil liberties, pornography, encryption software, tracking consumer behavior). What privacy should be allowed on the Internet? What is the difference between privacy and anonymity? What steps can you take to ensure your personal privacy online? How does that privacy relate to the various subjects of this book?
4. *Reorganize the Book.* There are different ways to categorize and organize information. Skim this book, examining the topics discussed. Create a new table of contents organized in a different way—a way that makes sense within your framework of understanding of human communication on the Internet.

Concepts for Analysis

We have provided concepts relevant to this chapter so that you can check your understanding. These concepts are worth dissecting in more detail. As a research option, examine one concept. Use our citations as a starting point for your study. We recommend that you search for refereed articles in an online database. Find at least three solid references you can read. Summarize, scrutinize, and present your analysis to other people in your course.

Questions to stimulate your inquiry: Do you agree with the ideas as we presented them? What other points of view should be considered? What details need to be added to fully understand the concept? What can you contribute through your personal analysis of the concept? What more do scholars need to know about the topic?

Blackboard is a popular course environment. http://www.blackboard.com/

Consciousness is awareness and perceptual understanding.

Convergence is a blending together. First, there is the convergence of technology. We have an increasingly difficult time distinguishing between our stereo, television, telephone, Internet service, and other media technology. Second, we have an increasingly difficult time distinguishing between our intra, inter, group, and mass communication modes. Convergence is the integration of various media and communication modes through a unified technology.

Course Compass is a course platform available to faculty and students through Pearson (Allyn and Bacon/Longman). Course Compass provides flexible tools and content resources that enable you to easily and efficiently customize Pearson Online Content to best suit your needs. The system provides course management and assessment tools. http://www.booksites.net/solutions/aboutcc.htm

Course environment is a computer package usually operated on a server, which provides an array of educational services, such as grading, testing, web links, course chat, and e-discussion. WebCT, Blackboard, and Course Compass are course environments available for this textbook's support.

Cultural homogenization is the domination of United States culture over other cultures, which many people believe is accomplished, in part, through the Internet.

Empirical research is a systematic investigation of the observable.

End users are the people who communicate via the Internet, in contrast to technical engineers who design and maintain computer hardware and software for communicating on the Internet.

Global village—entailing messages and forms of communication widely distributed—brings people around the world together through communication as a means to reducing cultural isolation and distance.

Interpersonal communication is face-to-face communication between two people. Internet discussion is more like interpersonal communication than like any other type of communication.

Intrapersonal communication is meaning-making with oneself, including imagined interactions, self-talk, journaling, and similar consultation with the self. We define intrapersonal communication as assigning meaning to stimuli and producing meaningful stimuli, regardless of whether those stimuli are verbal or nonverbal.

Mass communication is message transmission that generally occurs simultaneously and rapidly to a very large number of people. Mass media refers to technology that intermediates communication. Mass communication is often used interchangeably with mass media, although the word media refers to some kind of conveyor, means, or technology.

Meaning-making is the process of attaching significance or understanding during communication.

Paradoxical nature of Computer-Mediated Communication is the contradictory consequence of communication on the Internet. The strategies, values, and successes of communication on the Internet contain inconsistencies, opposing forces, and illogic.

WebCT is a popular course environment. http://www.webct.com/

Note

1. See paradox at: http://dictionary.cambridge.org/define.asp?key=paradox*1+0

2

Process of Human Communication on the Internet

> *When we look for the essence of a technology, we are engaging in speculation, but not in airy speculation. Our speculation involves where we plant our feet, who we are, and what we choose to be. Behind the development of every major technology lies a vision. The vision gives impetus to developers in the field even though the vision may not be clear, detailed, or even practical. The vision captures the essence of the technology and calls forth the cultural energy needed to propel it forward. Often a technological vision taps mythic consciousness and the religious side of human spirit.*
>
> —(Heim, 1999, p. 27)

We may not be able to unveil the mythic consciousness or cultural energy behind the Internet, but we have a sense of where we are heading and we want you to be looking along with us. Our exploration may lead, for example, to our cultural notions of democracy, commercialization, nation, meaning (real and unreal), or even isolation. We don't know, but we want to open the question: What happens to our communication because of the Internet?

We think good research is like good construction; creation requires a good dose of thinking and planning before you start cutting the material, gluing parts together, and banging in the nails. Of course, we also learn from building and collecting data without a strong foundation—usually we learn how mistaken we were in our understanding. People in the United States need digital literacy to succeed in our technological society (Gilster, 1997). We invite you to take part in this creation, to use your best thinking, to experience the Internet, and to critique the ideas you encounter here.

Focus

TO: Students

FROM: paradoxdoc@hotmail.com

RE: Overview: Process

Lenny and Joan elaborate on their ideas about the process of communication on the Internet. They discuss the importance of Internet literacy as an "appreciation of" course, representing a complicated approach to using and understanding the Internet. They challenge you to understand how people respond to changes in their communication lives, what they seek out, and what they avoid, and why. How would you explain the metamorphosis—changes of information, of people, of behaviors, of media, of society, of the world—caused by communication on the Internet?

Lenny and Joan ask you to consider mass communication concepts, such as technological determinism, which is the idea that the technology is the primary force that controls how individuals and society change. They ask you develop your own cyberliteracy, to consider the development of computers and the Internet, and to reach a common understanding of basic terms relevant for studying the Internet.

Lenny and Joan ask you to consider the informing and conversing nature of the Internet.

Lenny and Joan suggest you notice the political, cultural and economic forces that operate with the communication forces on the Internet, in addition to ideas about the diffusion of the innovation of the Internet, computer-mediated communication (CMC), language, time and space, and video technology.

As a way of viewing communication on the Internet, Lenny and Joan ask you to consider research methodologies and how that might affect your understanding of the Internet. And like most communication books, they offer a model of their perception of the process of human communication on the Internet.

Why is looking at human communication on the Internet important? Is it important? How does the Internet affect the way we interact? What do we mean when we propose that human communication on the Internet is paradoxical? What is the framework we adopt for thinking about communication on the Internet? At the end of this chapter, you should be able to answer the following questions.

1. What is the process of human communication on the Internet?
2. Is the Internet a mass medium?
3. What theoretical roots are particularly relevant to the study of human communication on the Internet?
4. What is your perception of communication on the Internet?

Developmental Perspective

Departments in colleges and universities across the nation are offering all kinds of computer courses. Some "turf protecting" has surfaced, as faculty attempt to define who should teach what courses. Previously, the majority of computer instruction has been in programming, data processing, and areas that relate well to science and mathematics. A relatively new area of study appropriate for the well-educated person of today is computer and Internet literacy. The purpose of such courses is not so much to teach computer operation as computer and Internet understanding. Internet literacy is not simply a skills course; Internet literacy is an "appreciation of" course, representing a complicated approach to using and understanding the medium (Gurak, 2001; Hawisher & Selfe, 1999).

As communication scholars, we want to understand how computers affect communication and how people use computers to communicate. We also want to understand how people respond to changes in their communication lives, what they seek out, what they avoid, and why.

We are fascinated by the multiple kinds of *metamorphosis* caused by the Internet. A metamorphosis is an evolutionary change. When a caterpillar changes into a butterfly, for example, there is a metamorphosis of form. In observing the Internet, we have observed more than just change, we have observed evolutionary change. The Internet causes metamorphosis: of information, of people, of behaviors, of media, of society (Fidler, 1997). Consider, for example, the idea of *technological determinism.* Technological determinism is the idea that the technology is the primary force that controls how individuals and society change (Winston, 1995). Do you think the technology changes the individual? Do you think the individual shapes the technology? We believe there are no simple answers to these questions because of the paradoxical nature of the Internet.

We believe educated people must understand the Internet's power on individuals, groups, societies, and the world. And we believe a course on human communication and the Internet—essential in today's communication world—fits well alongside such standards as interpersonal and mass communication courses. We consider this understanding to be a new form of literacy. Gurak (2001) explains that literacy in the online environment is not simply about reading and writing on a computer. Literacy today is about understanding the technology and how the process affects our lives, our thinking, and our interacting. She explains:

> . . . cyberliteracy is not purely a print literacy, nor is it purely an oral literacy. It is an electronic literacy—newly emerging in a new medium—that combines features of both print and the spoken word, and it does so in ways that change how we read, speak, think, and interact with others. Once we see that online texts are not exactly written *or* spoken, we begin to understand that cyberliteracy requires a special form of critical thinking. Communication in the online world is not quite like anything else. Written messages, such as letters (even when written on a computer), are usually created slowly and with reflection, allowing the writer to think and revise even as the document is chugging away at the printer. But electronic discourse encourages us to reply quickly, often in a more oral style: We blur the normally accepted distinctions (such as writing versus speaking) and conventions (such as punctuation and spelling). Normal rules about writing, editing, and revising a document do not make much sense in this environment. So it is not adequate simply to assume a peformative literacy stance and think that if we teach people to use computers, they will become "literate." Cyberliteracy (again noting Welch) is about *consciousness*. It is about taking a critical perspective on a technology that is radically transforming the world. (pp. 14–16)

Where Do We Start?

Perhaps we should start with the invention of the computer. Some people say the prelude to the computer goes back to seventeenth century Germany. In 1834, Englishman Charles Babbage invented an analytical machine, which some people consider the first computer. Howard Aiken and his associates developed the first modern operating computer during World War II. Commercial applications did not begin, however, until the 1950s. By the mid 1960s, new computer uses were expanding at a rapid rate. Management Information Systems (MIS) became an important concept in coordinating data for business and government in the 1960s.

At the risk of sounding like old fogies—okay, we are—we began studying communication nearly forty years ago. We can remember a computer system that took up multiple floors of a building, having less computing capability than the computers on our desks today. Back then, the formal, systematic study of interpersonal and mass communication was a new field. Berlo's book *The Process of Communication* (1960) gave us new ways of understanding the interactive nature of communication. McLuhan's book *Understanding Media: The Extensions of Man* (1964) popularized earlier works, which prompted us to view multimedia presentations and adaptations to different cognitive styles as new hope for adapting communication to all kinds of people. Stephenson's book *Play Theory of Mass Communication* (1967) explored the essence of play in communication, including mass communication, laying groundwork for the way scholars understand technology today (e.g., Pesce, 2000). In the 1960s, the way we imagined communication wasn't quite available to us yet, but we enjoyed exploring the means of communication for that time, and we cherished reading ideas of futurists about what might lie ahead. Would reality meet with the expectations of our imaginations?

The 1970s saw an explosion in database systems, telecommunications, time sharing, and personal computers (Walsh, 1981; Freed, 1999). As one scholar explained: "The computer is important in communication studies because its information processing abilities make it the major building block in the great array of developing communication systems" (Goldhamer, 1974, p. 94). That *systems* concept could include the personal, work, and family systems of the twenty-first century. Every new communication medium has affected those already in existence. The personal computer is what allowed individual access to the Internet. The introduction of personal home computers more than 20 years ago brought us closer to the way we envisioned communication. In the 1980s, when personal computers became less expensive, more user friendly, and more portable, personal computers began to proliferate rapidly. With their popularity for business, educational, and personal uses, scholars made positive and negative predictions about the future of human communication.

When the U.S. government set up a communication network between computer systems, they designed the system so they could communicate in time of emergency. We remember—not that long ago—when Lenny was experimenting with early, pre-Internet, communication systems and trying to persuade Joan to try them too. That network evolved into the Internet, which was opened for public use in the 1990s, with commercialization beginning in 1995. The Internet, that was what we were waiting for!

We're fortunate to have seen this communication metamorphosis happen during our professional career. We live during the most exciting time in history—a feeling we hope everyone has no matter what period of history they experience. And although we consider the Internet quite young and pliable, we cannot help but think about Paul Saffo's *30-Year Rule*. Saffo (1992) said that thirty years is about how long people take to completely embrace new technology. We believe the Internet has pushed up the 30-Year Rule to a 15-Year Rule. When our colleagues question whether to use the Internet in courses, we wonder why. The Internet is here, and the important question is not whether we will use the Internet, but: How do we want humans to communicate on the Internet?

The *Telecommunications Act of 1996* was a step to make the Internet accessible to everyone (Barnes, 2002, p. 295). Today, communication on the Internet is a pervasive activity in the United States. People refer to the importance of the information highway and the information explosion, while they open their homes, businesses, and schools to the world via the Internet. During the national emergency of September 11, 2001, for example, millions of people turned to the Internet to find information and to connect to family and friends. When telephone systems failed, the Internet worked as it was originally intended.

The Internet has provided a new communication medium. One question we must raise is: What sort of medium is the Internet? Is this a mass medium of communication with its attendant characteristics or a personal and interpersonal medium? However we answer this question, we do note that different people use the Internet differently; hence, we can expect a variety of user effects. We recognize that the Internet does act as a vehicle in the human communication process, but that fact tells us little about the process itself. The Internet does convey information. We will make a distinction between information and communication.

Communication scholars have studied communication *effects* only in relatively recent years—roughly since the 1930s. Many people blamed and credited mass media, for example, with diversified individual and social effects. Increased violence from television viewing is a widely accepted—but difficult to prove—social consequence. There are several reasons for inconclusive mass communication research: (a) the interaction of many variables, (b) the heterogeneous nature of mass communication, (c) the difficulty in demonstrating direct cause-and-effect relationships, and (d) the variety of ways in which people use the media. The complex nature of the Internet, *intervening variables* (i.e., variables that operate between what seems to be the cause and effect), and individualized responses also make scholars reluctant to suggest simple, straight-line cause-and-effect relationships regarding the Internet and communication effects. Research that reports such specific Internet effects does not exist and would be suspect if it did.

You may be someone who knows much about computers and the Internet, or you may be someone who has little knowledge about communication on the Internet. This book is designed to raise for discussion your beliefs about communication on the Internet, no matter what your current knowledge and use levels may be. So, before we proceed, we need to be clear about terms so that we all have the same definitions for words used in this book.

The *Internet*—also called the *Net*—is an electronic communications system that connects computer networks and organizational computer facilities around the globe.

The *World Wide Web*—also called the *Web* or *WWW*—is a part of the Internet designed to provide easier navigation of that network.[2]

Computer-mediated communication (CMC) is human interaction through computer technology.

Human-computer interaction (HCI) is communication between a person and a computer (the software, a computer robot).

The World Wide Web actually refers to the way people use the Internet, that is, how we use software such as Netscape or Internet Explorer, which allow us to see graphics and text, hear sound, read and write messages, and click on links. We think of the Internet as the infrastructure of computers, codes, optic fibers, and wireless waves, but the WWW is the content that is characterized by pointing and clicking on Web pages. While these explanations may not be 100 percent true, these definitions help us to conceptualize this amorphous technology. And that perception is a great part of the challenge to understanding computer-mediated and Internet-mediated communication. Computer-mediated communication (CMC) is communication in which a computer mediates or facilitates the interplay between people. We consider all human-computer interaction a type of computer-mediated communication because in reality, people are still communicating. When you play chess with a computer program, you are actually playing with the people who wrote the program. In other words, CMC is communication when a computer is used.[3] Of course, you could pick up a computer and purposely drop the machine on someone's foot, but that is not what we mean by the term computer-mediated communication, although it is communication involving a computer. And yes, writing a note on your computer, printing the message, and handing it to someone to read also fits this definition of CMC.

By Internet-mediated communication or communication on the Internet, we mean the processes of encoding and decoding messages transmitted on the Internet. Communication on the Internet may use wired or wireless means and may involve the convergence of various electronic devices. Thus, our focus will be on CMC on the Internet. So, when you use a computer to send someone an electronic message, for example, we consider that to be computer-mediated communication. If you sent that message over the Internet and the message was decoded by a person(s), then computer-mediated communication occurred, even if the receiver of the message is sitting right next to you, even if the receiver is you yourself, and even if the receiver is some unintended receiver, that process is CMC.

Common examples of CMC are reading e-mail and looking at web sites, as well as reading and writing to electronic discussion groups, or getting and reading online journals. Notice that while each of these examples is an instance of CMC, each is quite unique in communication terms. E-mail is often sent to an individual by an individual, often people with a personal history; web sites are often displayed for public consumption, although one may imagine the purpose of the web site and the audience intended; typically, discussion groups involve messages for those who share an interest; and online journals are likely to be (professional) organizations providing professional materials for professionals, much like paper journals do. Each has its own communication character.

Cyberliteracy

These ideas bear repeating:

> Cyberliteracy is not purely a print literacy, nor is Internet literacy purely an oral literacy. Cyberliteracy is an electronic literacy—newly emerging in a new medium—that combines features of both print and the spoken word, and the medium does so in ways that change how we read, speak, think, and interact with others. Once we see that online texts are not exactly written or spoken, we begin to understand that cyberliteracy requires a special form of critical thinking. Communication in the online world is not quite like anything else. Written messages, such as letters (even when written on a computer), are usually created slowly and with reflection, allowing the writer to think and revise even as the document is chugging away at the printer. But electronic *discourse*—talking, conversing, interacting—encourages us to reply quickly, often in a more oral style. In discourse, we blur the normally accepted distinctions (such as writing versus speaking) and conventions (such as punctuation and spelling). Normal rules about writing, editing, and revising a document do not make much sense in this environment. (Welch, 1999)

We agree that cyberliteracy is about consciousness. Consciousness—mental awareness—is about taking a critical perspective on a technology that is radically transforming the world.

You will notice that the word Internet is capitalized. While this point may seem minor, the capitalization indicates that Internet is a proper noun, not just a thing. The Internet is not simply a body of data on a technical medium: *The Internet is a human activity.* We can liken this distinction to "language" and "the use of language." Language can be studied as a thing, dissected, and laid out. Using language is dependent on the wide range of human needs, emotions, motives, contexts and so on. Language and language use are not the same thing. Likewise, CMC is a human activity, much like having a conversation is the use of language, or language in use. With its great affinity to conversing, interpersonal communication over the Internet often enacts the subtle features of conversation in which we construct reality with words and language (and conversational) structures. In interpersonal communication on the Internet, we must work out identities, status and power differences, and establish boundaries through negotiation or break down boundaries.

The field of communication studies has a tradition of examining the effects of mediated communication. As we explained, straight-line effects are difficult to establish. Because effects are difficult to determine, we often talk of associations. An effect is a direct result of media use. An *association* is a behavior linked or paired with media use. For example, heavy use of violent video games is associated with violent behavior. We can't say the violent behavior is an effect, nor can we say the violent games cause violent behavior, because people who are already prone to violent behavior may also be prone to play violent video games. Thus, we can say that violent behavior is *associated* with heavy use of violent video games.

A *context* is a place, an environment, a situation, or a conceptual framework in which communication takes place. Contexts typically involve underlying understandings such as those we find in relationships, say in the family or the workplace or a specific relationship, or even gendered patterns. But communication scholars often examine contexts as a level of communication, by looking at interpersonal, public, and mass communication. Such dimensions define the medium in communication terms. Typically, the definitions of television, radio, and other forms of mass communication are that they involve scientific methods through which messages are transmitted or conveyed. The term mass communication is often used interchangeably with mass media, although the word *media* refers to some kind of conveyor, means, or technology. The word media is plural of medium, and although we refer to the Internet as "a medium," we realize that multiple complex communication media are involved in Internet interactions.

In studying communication on the Internet, levels and media distinctions no longer make good sense. Modes and means become blurred and entangled. Modern human communication is so convergent and complex that mediated communication can be broadly interpreted to include features of face-to-face communication, writing, spoken language, culture, mind, and mass communication. Consider these instances of Internet use:

- The Internet enables a person to mentally escape while using Real Audio to listen to meditation guidance.
- The Internet enables a person to have intimate words with a lover.
- The Internet enables a face-to-face videoconference between two business associates.
- The Internet enables a cancer survivor to establish a support group and web site for people all around the globe.
- The Internet enables a student to find information for a research essay.

In these examples, you can see how the Internet enables communication at any level: intrapersonal, interpersonal, public, and mass communication. One scholar views communication on the Internet in levels: intrapersonal, interpersonal, and transpersonal (Gackenback, 1999). We contend that these distinctions are artificial. This differentiation may work for methodological reasons, but breaks down for theoretical reasons. Although a psychologist might see intrapersonal communication as synonymous with the self, a communication scholar sees how interrelated all elements of communication are. The Internet provides a way to examine how intrapersonal communication is involved in all elements of human communication. Because all levels of communication exist on the Internet, the distinctions are blurred. Thus, we have decided to look at contexts of human communication on the Internet, recognizing that intrapersonal, interpersonal, public, and mass communication may be involved simultaneously.

- How you will work.
- How you will carry on relationships.
- How you will spend your leisure time.
- How you will approach education.
- How you will acquire information of all sorts.
- How you will view yourself and others.

Perhaps an example of communication over the Internet in the business context will help to bring home these two points: (a) that the levels of communication converge, and at the same time, (b) communication on the Internet is both affected by communication variables, and affects the face-to-face world of communication itself. Shockley-Zalabak (2002), writing about virtual teams operating in the work context, finds that characteristics of virtual communication change the way businesses handle work, characteristics such as the speed of the communication online, the amount of data that can be shared, the reduced cost of communication, the increased ability to communicate text, visual, sound, video, and graphics in a multimedia environment, connection to large numbers of computers (connectivity), and the convergence of communication and computing. The ability of computer systems to store and retrieve information and to organize information in databases that can be shared, all adds to the movement toward team work. Change in how people work affects the way they are managed, the way they interact with one another, the way they interact with clients, who they work with, the projects they work on, when they work, their perception of their place in their team, and how they gain feedback for their work.

The *medium* affects the way we communicate, sometimes improving our ability to communicate in certain ways and sometimes making the communication more difficult. To use the Internet to communicate via computers on the World Wide Web has implications regarding with whom you communicate, how you communicate with them, and when you communicate. Who, how, and when have implications for the ways we conduct our lives.

Why Study Human Communication on the Internet?

We have no final destination planned for our expedition. By that we mean that we are serious about your part in exploring the nature of human communication on the Internet. Scholars have not yet found the best way to describe communication on the Internet. Communication on the Internet is still open to new understandings. We can apply research and theory to our exploration—and we will introduce you to some theory—but there is much more available. We give you an interpersonal communication perspective.

Please join us in this adventure as we converse about human communication on the Internet.

We have said that we are struck by the paradoxes and oppositions that lie within communication. We are intrigued by the effortful or conscious way in which we try to make sense of symbols presented on the Internet, as well as the opportunity to spend time mindfully wandering about the Web. We are struck by the potential for the Internet to enter into so many parts of our communication lives. We cannot ignore our own experiences with the Internet. As academics, we read the academic literature available online about the Internet, and we write some of that scholarship. We quote from the research literature in our work. We edit online publications. We interact with our students, with one another, and with colleagues. Our collaborations include academics in professional organizations, our publisher, the National Communication Association, the union, and our universities. As members of the Internet generation, we access the news and entertainment online. We both design and maintain a variety of web sites. We visit online, we stay in touch with friends and relatives, on and on. We are praised for some our online academic work and punished for it as well (Shedletsky & Aitken, 2001).

Because of our varied experiences, we consider this book more than an academic exercise. We use the Internet to explore ourselves, each other, other people, our society, and the world. That exploration is an adventure. As communication scholars and as human beings, we agree with the statement: "The impact of digital technology is truly transforming all forms of human communication" (Pavlik, 1998, p. 2).

The Internet is a social and psychological phenomenon, which has been studied by scholars in various fields (e.g., Hobman, Bordia, Irmer, & Chang, 2002; Kraut, Kiesler, Boneva, Cummings, Helgeson, & Crawford, 2002; McMurdo, 1995). Early work focused on the psychological and social nature of the Internet (Reid, 1991), but our focus is on the communication. Understanding communication on the Internet puts people in a position to perform in the "real" world and to maximize their performance. Just as importantly, and maybe more importantly, understanding communication on the Internet heightens our awareness of the world in which we make meaning. Because this book explores how Internet communication matters, we seek to stimulate your thinking about the psychological and social communication implications of the World Wide Web. In other words, within a framework for thinking about computer-mediated communication (CMC), you can learn how to enhance your chances to operate well, to flexibly respond to particular situations, and to create extensions of your knowledge. But we believe that this book ought to help us do more than to maximize performance on the Internet; this text should help us understand better who we are as humans, what we strive for, how we evolve as communicators.

At the same time, Internet-mediated communication is a new and unique area of study, which creates special constraints on our ability to make sense of that communication. If James Carey is right about the paradoxical nature of communication on the Internet, and we think he is, then making sense of this new way of communicating is particularly confusing. This text builds upon research, by theorizing, observing, speculating, and exploring. Given the state of the art of research on CMC, you will find much of what we have to report is speculation.

Communication Media

Communication scholars pay attention to the ways in which a particular medium influences the communication process, by examining the dimensions of communication that are enhanced or diminished or even hidden by a medium (Table 2.1).

Language and Writing

One can compare computer technology to other communication media. To begin with, communication on the Internet incorporates two fundamental media or forms of human communication: (a) language and (b) writing. Media scholars have recognized that language and its derived form of writing have transformed human experience (Baron, 2000; Fidler, 1997). Language offers humans the ability to represent ideas both for thinking and for expressing. People conversing on the Internet have developed their own language conventions, including acronyms (e.g., LOL meaning laughing out loud), graphic accents (emoticons such as ☹ to indicate sad), ASCII visuals which are completely textbased, and verbverbing (such as hughug a practice of action used in MUDs, or multi-used dimensions, which are primarily game-playing or educational environments (Shah & Romine, 1995)).

Writing offers humans the opportunity to record ideas, to preserve them, to transport ideas, and to distribute ideas to many people. Seen in this light, media allow humans to organize themselves in new ways, create, maintain, and transform their perceptions. For instance, something as simple as our ability to send one another an e-mail message on the Internet can influence communication simply because we do not both need to be present at the same time. First, we can edit what we say and take our time in putting it together. Second, we can take our time in thinking about the other's message. Third, we may feel safe not seeing the other or being seen; our message is preserved in written form. Fourth, we do not need to respond immediately to the other. Finally, we can leave the "conversation" at any time. Dimensions of the communication medium matter.

The ability to cross time and space with our communication is tied to our ability to send and receive information, and to express our beliefs. We can see that wireless media in the form of broadcasting would heighten our ability to extend our ideas in time and space. With regard to the Internet, we see that much of what has gone before is incorporated into the media of the Internet.

TABLE 2.1 *The Medium Can Enhance or Diminish Communication*

Enhances Communication	*Diminishes Communication*
Writing enhances time to edit.	*Writing* diminishes speed of interaction.
Reading enhances time to interpret.	*Reading* diminishes two-way communication.
Film enhances movement through time and space.	*Film* diminishes flexibility and is quite linear once the message begins.

Visual Medium

Like film and television, the Internet is sight- or visually oriented. The computer has many characteristics of mass communication media. Mass media generally are able to transmit messages simultaneously, at a low unit cost to many people, using high-speed techniques of reproduction and distribution. The Internet permits the transmission of information from a source to a large number of people. The Internet is a rapid communication vehicle, and affects society as a whole.

Let's consider some examples of that idea. Film began as a pervasive mass entertainment medium, with information forming a byproduct. Over the years, probably as a result of television, the mass appeal of film has changed so that it has evolved into an art and entertainment youth medium. The Internet also is popular as an entertainment youth medium. Television began as an entertainment mass medium that proved to have educational and informational values. Now television has undergone massive changes and diversity through the new technology of satellites, DVDs, cable, and the Internet. The Internet was designed to be an emergency communication system, but has evolved into an interpersonal, entertainment, economic, and information medium, often used to improve professionalism and efficiency. Although the Internet may not replace television, it may converge with television to create significant changes in the medium of television. While the television helps people kill time effortlessly—as can the Internet—the Internet can help people learn, create, and communicate effortlessly.

The Internet can use video technology. Video technology can be absorbing and engrossing. The Internet can be so absorbing and engrossing because it uses video technology and because of its interaction capability. Research has suggested that children prefer computers to television because of their interactivity. Perhaps the interactive nature of the Internet, in contrast to the passive nature of television, for example, has lead to more positive intellectual uses. The new attraction has lead to the change from couch potato to computer potato. But while being a couch potato is actually a mind-numbing activity, being a computer potato is a mind-stimulating activity.

At the same time that we consider the Internet as mass communication, we can talk about the Internet as an instrument of interpersonal communication. Interpersonal communication refers to messages communicated between two people. Indeed, the leading use of the Internet is to send and receive e-mail. You probably have heard of romantic relationships springing up from online encounters. People meet online. They talk; they flirt; they rage; they express themselves to one another through the Internet. Friendships are created and maintained online. Students and teachers interact electronically in groups and dyads.

If you were to think of ordinary, everyday communication—e.g., talking to a friend face-to-face—as satisfying certain basic needs or motives, then it is not difficult to apply the same core concepts to communication on the Internet: Communicating to satisfy our social needs for (a) pleasure, (b) affection, (c) inclusion, (d) escape, (e) relaxation, and (f) control (Rubin, Perse, & Barbato, 1988). No doubt we could argue about the exact list and whether these six are the whole list, but the larger point is that these sorts of communication-related concepts would seem to apply to communication on the Internet.

You might want to add additional motives for communicating, such as developing a sense of who we are, identity, or getting practical needs met, or just connecting (Adler, Rosenfeld, & Proctor II, 2001). As we take a closer look, we may find that there are additional needs, motives, or dimensions of experience that are enacted as we communicate on the Internet.

We want you to consider how the medium of the Internet shapes communication. The Internet doesn't change the fact that you communicate with others, but the Internet enhances or *intensifies* certain ways that you communicate. These ideas (see Table 2.2) are elaborated and challenged in this book.

TABLE 2.2 *Communication Intensification*

Internet Enhances Communication	*Internet Diminishes Communication*	*Internet Shapes Communication*
• Intensifies mental aspects of communication.	• Disembodies communication.	• Shifts attention to communication dimensions.
• Increases speed of communication when people are apart.	• Can confuse meaning because there are few or no nonverbals to help you interpret the message.	• Empowers the self.
• Connects family and friends.		• Structures cognitive processing.
• Provides opportunity for storytelling.	• Polarizes people.	• Enables new learning and understandings.
• Maintains better relationships.	• Creates educational controversies.	• Creates new ways in which knowledge is created, expressed, and distributed.
• Brings people together.	• Intensifies distrust.	
• Contributes to the development of community through e-groups.	• Blurs boundaries between work and the personal.	• Creates new freedoms.
	• Blurs boundaries between fantasy and reality.	• Creates interdependence.
• Intensifies change while life remains the same.	• Diminishes a sense of safety through games of destruction.	• Creates new relationships.
• Gives an outlet for facing fear.	• Creates cognitive dissonance.	• Empowers people through support groups.
• Solves educational problems.	• Gives voice to hatred.	• Intensifies expression of feelings.
• Intensifies a sense of connection.	• Gives empathy to violence.	• Intensifies imagination.
• Intensifies learning.	• Diminishes self-confidence through embarrassment over hoaxes, rumors, and myths.	• Permits personal experimentation.
• Enhances concentration;		• Causes an interplay with the self or other.
• Enhances play.		
• Connects human talents.	• Enables personal addiction to the Internet.	• Causes interplay across age, gender, and ethnicity.
• Resolves the inner/outer dichotomy.	• Increases fears of hatemongers, hate sites, stalking, and murder.	• Creates metaphoric meaning.
• Enhances collaboration.	• Facilitates cyberterrorism.	
• Provides interactivity.	• Diminishes privacy.	
• Increases freedom and flexibility.	• Threatens freedoms.	

Models of Communication

Theorists have used models to help conceptualize and explain various processes, including the process of human communication. Modeling has helped people better understand and work with complicated ideas and the abstractions of processes, operations, and systems. The many communication models in our field suggest a variety of ways of conceptualizing the communication process. Although no single model has completely described all people in all situations, some models are more effective than others. We will discuss a general model and propose an intrapersonal model of communication to apply to Internet communication.

Most communication models include a *sender* of verbal and nonverbal *messages,* which is then sent through a *channel* to a *receiver.* The receiver must *decode* the message and gain some mental conception, preferably one that is similar to the source's original meaning. A variety of types of *noise* obstruct the success of the process. The noise is anything that interferes with transmission and can include internal and external influences. Upon receipt of the message, the receiver sends *feedback,* or a response, to the source. The receiver must also *encode* the feedback and send it by a channel to be decoded by the source. Shannon and Weaver, writing in the 1940s, presented a model of communication that fits this description of sender and receiver, with a signal traveling through a channel. Current day theorists look back on the Shannon and Weaver (1999) model as a linear representation of the communication process, sometimes referred to as a transmission model. That is, it is seen as a step-by-step model as opposed to a more holistic or simultaneous process. As a transmission model, it depicts communication as a process of moving or transmitting information from one point to another, as if in a straight line. Alternatively, the simultaneous view of communication is referred to as transactional. Most current-day communication models picture the human communication process as transactional, circular rather than linear, without clear-cut starting and ending points.

There can be little doubt that the linear model of communication has deep roots in our cultural view of communication, and that that view continues to play a role in the perception of communication, both online and off. It is a view that highlights moving things—information—from one place to another; quantifying how much is moved or stored; and likening the process of communication to the process of transportation through space. But as Jones (1997b) says, "Having information and knowing what it means are entirely separate domains" (p. 4).

Let's relate communication on the Internet to this generalized model of the communication process. Imagine an individual decides to send email to a friend. We can think of the sender as the source of the message, the computer and the Internet as the channel and the friend as the receiver. Imagine that Lenny decides to send e-mail to Joan. The receiver, Joan, would decode the message and send feedback in some form (including silence). Of course, as you would know if you have sent e-mail, this example is an enormously simplified description of the e-mail event because we have left out the intrapersonal processes of deciding to send e-mail, the construction of the message in terms of word choice, syntax, style, and propositional content. What is the intention behind the message and the social action communicated?

Imagine Lenny is wondering how Joan is doing (see Figure 2.1). Why would he be wondering that? For one thing, he hasn't heard from her in a while and so he senses that

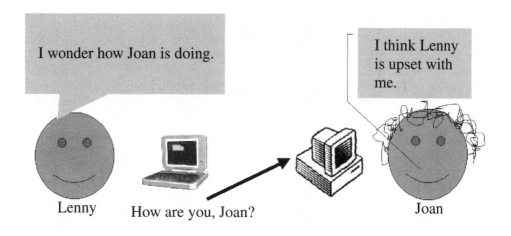

FIGURE 2.1 *Meaning Communicated via the Internet*

maybe she is having a tough time and is reluctant to talk. Perhaps Lenny knows Joan was worried about her injured pet. So, he writes to her and asks, "How are you, Joan?" Joan receives the e-mail and wonders if Lenny is upset with her. Why does she interpret his message as possibly signifying that he is upset with her? Maybe because that is all he said, but he usually tells her what is going on with him. Maybe Joan feels that Lenny at least discusses something about their latest work project when he writes, so his lack of communication suggests a problem. In reality, he only meant the question literally because Lenny knew of the injured pet, but was in a rush and could only send a brief e-mail.

At this point, you may be wondering if we are talking about (a) intrapersonal communication within an individual's mind, (b) interpersonal communication, (c) mass communication, or (d) some new form of communication. Let's respond to this confusion by recognizing the innovativeness of communication on the Internet. The Internet may require new categories, perhaps ones that we are not familiar with yet. We also recognize that e-mail messages often are difficult to interpret.

Messages received on the Internet are notoriously ambiguous. Like Joan and Lenny, many people have had the experience of not being sure just how to interpret an e-mail message. When meaning is ambiguous, that is when we tend to become aware of *attributing meaning*.

Social behaviors that are not easily understood are subject to attribution-making as people try to assess the cause, and ultimately the meaning, of the observed actions (Heider, 1958). The ambiguity of certain behaviors allows people to decide whether the cause of the action was internal or external to the other, whether it reflected a stable characteristic of the person or the situation, whether it was a specific or global cause, and whether the other acted intentionally. Each of these choices helps determine what meaning can be taken from the other's behavior because each leads to different interpretations and evaluations of the action (Manusov, 1995).

Reflection

When you look at the mirror of the Internet, what do you see? A reflection of yourself? Your world? What do you already know about the nature of human communication on the Internet? We—Lenny and Joan—are concerned with the ways in which communication on the Internet influences who we are and how we live, how we see ourselves, and how we think. The expanding and pervasive nature of the Internet affects all Americans. An understanding of human communication models can provide a way for you to increase your comprehension of the process. A better understanding of how humans communicate on the Internet can help you shape your future in the ways you desire.

We wrote this book from the point of view of people communicating, not people computing; we are not computer scientists nor are we hardware or software experts. We are experts about the communication process and "users" like many of you. That doesn't mean that we take advantage of people ☺. That means that we haven't spent much time opening up the box covering our computers or learning about programming; we are end users of the Internet. We have thought about, read about, and systematically conducted research about the Internet. Our interest is in human communication as a social and psychological interactive phenomenon. We are interested in how this new way of communicating matters, so we want to stimulate your thinking about the psychological and social implications of human communication on the Internet.

You may wonder why anyone would want to explore all these ideas about the nature of communication on the Internet as opposed to simply discussing how to communicate. We believe that with a framework for conceptualizing Internet-mediated communication, you improve your chances of communicating effectively, of flexibly responding to particular events and contexts, of feeling empathy in your relationships, and of behaving responsibly.

The Internet can shape how you communicate with yourself and others. Perhaps the Internet transcends the individual and interpersonal. Mediated environments and virtual reality do affect cognitive processing, perceptions of reality, and communication with the self and others. We like the metaphor of the Internet as a global brain. Perhaps the Internet functions more as a global soul, however, which intensifies your interplay within the self and other people. What is your interplay through the Internet?

Case for Discussion

In your course discussion group, give your analysis and perceptions on the following.

We began studying computer-mediated communication some twenty years ago. Over the years, we have come to use the Internet in many ways. We shop online, create web pages, provide community service, teach, work, and play online. We have used online strategies to teach more than a dozen different courses in communication studies. We conduct research, access information, and learn online.

We developed a successful work relationship online before we ever met face-to-face. In our case, we edited a book by sending files back and forth on the Internet and talking about our work on the Internet. We discovered a similar theoretical bent with an interest in the subjective. We still work via the Internet, although we have met face-to-face several times over the past several years. Our relationship, however, is dependent on our interplay through the Internet.

We interviewed and surveyed over 100 people to gather their perceptions of computers and the Internet. We also participated in close to 100 educational and non-educational online discussion groups as a way to gather ideas. Make no mistake, we were not always working, we often used the Internet for sheer pleasure. And to us, the lines between work and pleasure—business and personal—become blurred where the Internet is involved.

1. How might our Internet experiences influence how we attribute meaning to the Internet?
2. How does human communication on the web work in your personal experience?
3. What elements of human communication are intensified by communication via Internet technology?

Internet Investigation

1. *Web Resources.* There is a Course Compass that accompanies this book. Explore this now. While reading this book, immerse yourself in the Internet. If you are not already an avid Internet user, give it a try. Join some new e-mail or message board discussion groups. If you have never used chat or instant messaging, try it. Send greetings to the people you love. Find out what databases are available through your college and local public library, and explore each of them. Use the Internet to learn in a holistic way.
2. *References.* Examine the reference list at the end of the book. Look up and read one article written by Aitken or Shedletsky or another author cited in this chapter. How might that research affect the authors' perceptions of human communication on the Internet?
3. *History of the Internet.* Usually when you investigate historical events, you need to research developments of long ago. The history of the Internet is quite short and amazingly complicated for the brief span of time since the Internet began. Investigate the history of the Internet. Take a generalist approach to the history of the Internet or select a specific topic—e.g., packet switching, ARPANET, the development of PCs, advances in wireless computing, portable devices, embedded computing—and write a report you can share with others in your course. How has the historical development of these technologies affected communication?
4. *Ideas of Marshall McLuhan.* Investigate the concepts of Marshall McLuhan. What ideas are helpful in analyzing human communication on the Internet? Some places to start your investigation include online sources: http://www.mcluhan.ca/ and http://www.mcluhan.utoronto.ca. Paper resources are also good starting points: McLuhan, M. (1964). *Understanding media: The extensions of man.* New York: McGraw-Hill, or DeKerckhove, D. (1997). *Connected intelligence: The arrival of the Web society.* Toronto: Somerville House.

5. *Language on the Internet.* Language sometimes takes on unique characteristics when used on the Internet. Investigate the work of a scholar who has written about language and the Internet (e.g., Gozzi, 1999; Lanham, 1993; Lee, 1996; Ong, 1982). What is something you have noticed about language on the Internet? Explain your observation in the framework of published scholarship.

Concepts for Analysis

We have provided concepts relevant to this chapter so that you can check your understanding. These concepts are worth dissecting in more detail. As a research option, examine one concept. Use our citations as a starting point for your study. We recommend that you search for refereed articles in an online database. Find at least three solid references you can read. Summarize, scrutinize, and present your analysis to other people in your course.

Questions to stimulate your inquiry: Do you agree with the ideas as we presented them? What other points of view should be considered? What details need to be added to fully understand the concept? What can you contribute through your personal analysis of the concept? What more do scholars need to know about the topic?

30-Year Rule. Saffo said that 30 years is about how long people take to completely embrace and use a new technology.

An *association* is a link or coordination of a behavior which is paired with something else. For example, violent video games are associated with violent behavior. In other words, an association is a relationship between Internet use and various activities, behaviors, and experiences. Associations are not necessarily effects.

Attributing meaning refers to giving or assigning meaning to something, understanding it a certain way.

Channels are our senses, and provide the ways through which we send a message. We convey a message through sound, sight, smell, touch, and taste.

Computer-mediated communication (CMC) is human interaction through computer technology.

Consciousness is awareness and perceptual understanding.

Context is the situated meaning, environment, timing, occasion, or physical place that affects the way communication happens—the way meaning is assigned. The Internet alters the context. In some ways, but not all ways, the Internet transcends the time and place of communication.

Cyberliteracy is the knowledge and expertise needed to effectively use and understand the Internet.

Decoding is the process of analyzing a message in order to obtain meaning from it. In other words, decoding is when the individual makes sense out of the message by attending to, processing, analyzing, and remembering the message.

Discourse is the process of conversing, interacting, and talking.

Effects are direct results of mediated communication.

Encoding is putting a message into language and other symbols so that it can be sent to another person (or even to one's self). Encoding is the process of translating an idea into language, and therefore, a message.

Feedback is the communication response. As a public speaker, you'll want to observe audience reaction, facial expressions, head nodding or shaking, restlessness, and any cues that give you information about their response to your message.

Human-computer interaction (HCI) is communication between a person and a computer (the software, a computer robot).

Intensification is the ability of the Internet to enhance, deepen, or increase elements of communication. Walther (1996), for example, calls communication on the Internet "hyperpersonal."

Internet—also called the *Net*—is an electronic communications system that connects computer networks and organizational computer facilities around the globe.

Intervening variables are variables that operate between what seems to be the cause and effect. In a sense, they challenge the researcher/theoretician to understand how a phenomenon works.

A *medium* is the way communication is conveyed—e.g., computer, television—which affects the way we communicate, sometimes improving our ability to communicate in certain ways and sometimes making the communication more difficult.

The *message* is the encoded idea or understanding that the speaker shares with the audience.

Metamorphosis is an evolutionary change. When a caterpillar changes into a butterfly, for example, there is a metamorphosis of form. In observing the Internet, we have observed more than just change, we have observed evolutionary change. The Internet causes metamorphosis of information, of people, of behaviors, of media, and of society (Fidler, 1997).

Noise is a distraction during a communication situation. Noise can be actual noise, a thought, a visual distraction, or other influences that break your focus. Noise is anything that interferes with transmission and can include internal and external influences.

The *receiver* is the one who is given a message, such as the individual who receives an e-mail.

The *sender* is the source of a message, the person who sends an email for example.

Syntax is the way a language fits together—structure of utterances in terms of nouns and verbs, clauses, interrogatives, imperatives, and grammar.

A *system* is an integrated, interactive, working structure, such as a computer network or the interpersonal organization of a particular family.

Technological determinism is the idea that technology is the primary force that controls how individuals and society change (Winston, 1995).

World Wide Web—also called the *Web* or *WWW*—is a part of the Internet designed to provide easier navigation of that network.

Note

2. http://www.m-w.com/dictionary.htm
3. http://www.ucfv.bc.ca/aded/seeds.htm#SEC2

3

Tensions of Communication on the Internet

> *Modern communication is developing along two fronts, each creating its own parallel and yet contradictory phenomena. On the one hand, there is the promise of the globalization of personal communications, with its potential to empower individuals far beyond the wildest dreams of the speakers who shouted their messages from mountaintops. On the other hand, there is the reality of globalization: continuing centralization of mass communications, with a few players (often international corporations) controlling the chokepoints, leaving the overwhelming majority of the world's people increasingly marginalized on the periphery.*
>
> —(Mowlana, 1995, p. 2)

Focus

```
TO: Students

FROM: paradoxdoc@hotmail.com

RE: Overview: Tensions

Lenny and Joan apply the idea of dialectical tensions to
the paradoxes of the Internet. They ask you to consider
negatives about the Internet, in addition to the
positives. They discuss globalization as an example of
paradox of the Internet.
```

Lenny and Joan discuss the term "cyberspace" as a metaphor for some sort of imagined environment or psychological experience. The Internet communication experience is a contradictory pull between the hard stuff of ordinary reality, like talking to people, for instance, and the "unreal" stuff of the virtual world, like interacting with an imagined person. Much of communication on the Internet is about change, including changes in the self, the perception of the world, the way we communicate, and with whom we communicate. Regardless of how Internet use evolves, people who live in the United States are using the Internet more and more.

The diffusion of innovations happens in a particular way, as part of a process of social change. Diffusion of Internet use has been influenced by social, economic, political, educational, and personal factors.

Lenny and Joan ask you to consider the theoretical roots that support Internet research, including dialectical tensions, linguistic study of discourse, and the inner-outer dichotomy. The humanist examines how the individual makes sense of the world from personal perspectives and conceptions. The research approaches of transformational grammar, semiotics, discourse analysis, and ethnomethodology reject operationalization and its associated requirement for all observable variables. Additionally, these research perspectives take into account underlying structures of the mind.

The behavioralist examines human communication as essentially rooted in observable behavior, as witnessed in defining all variables studied in observable terms or operationalization.

At the end of this chapter, you should be able to answer the following questions.

1. Explain the concept of globalization as it relates to the Internet.
2. How do you envision cyberspace?
3. Does the user of the Internet feel a trust that is absent with other mass media?
4. In what ways is communication on the Internet more conversing than informing?
5. What is the process of adoption and diffusion of technology?

6. How do dialectical tensions relate to Internet paradoxes?
7. What are theoretical differences between the behavioralist and the humanist that could shape your understanding of the communication?

Globalization

If we are not careful, you might have the impression that we have been swept along with the hype of the day. Perhaps you think we have made our minds up about the Internet, that we are true believers, and good citizens of the new communication order. Not so. We find more scholars who praise the Internet than those who are concerned about its ill effects, but there are critics out there (e.g., Postman, 1986; Postman, 1993; Postman & Paglia, 1999). As with all new forms of communication, the Internet has enthusiasts and skeptics. The enthusiasts emphasize the democratic potential of the Internet, including technology's ability to add to the citizen's direct participation in political decision making. Opponents of this view speak of the problems caused by free access to information, lack of centralized control, and increased globalization caused by the Internet.

Globalization is the process whereby nations become increasingly dependent regarding military intervention, finances, and communication. Some people believe the process of globalization is natural and positive. Other people believe globalization is negative, primarily because globalization destroys individual cultures and westernizes the world. In the District of Columbia in 2002, for example, 600 people were arrested during their demonstrations against the globalization effects caused by the international monetary fund and the world bank.

The mass media are credited by some and blamed by others for contributing to globalization in ways that interfere with individual ethnic and national traditions. Although many people believe the Internet contributes to globalization, other people believe the opposite happens. These people believe the Internet gives individual cultures a communication vehicle to express themselves to the world. Globalization interferes with customs and mores of individual cultures, which makes globalization a controversial effect of media. What is the role of the Internet in globalization? Do knowledge and ideas on the Internet take on a unified voice or more individualized voices? As it goes with the many paradoxes of the Internet, we find truth in all sides.

Cyberspace

Some people use the term "cyberspace" to refer to the Internet, or "being on the Internet." That almost sounds like being on some sort of mind-altering drug. A person might say "you are in cyberspace." But surely you are not in any space other than in your home or at work or somewhere near a computer or maybe even an electronic device connected to the Internet, like a cell phone. *Cyberspace* must be a metaphor for some sort of imagined environment or psychological experience.

Benjamin Woolley (1999), tracing the meaning of "cyberspace," takes us through a variety of views on communicating on the Internet. Woolley helps us to think about just how this new way of communicating influences our view of the self and the world. There may be hints of meaning in the very terms used: cyberspace, virtual worlds, virtual space, the information sphere, and others. One begins to see in the terms that refer to the Internet or being on the Internet. The Internet is a contradictory pull between the hard stuff of ordinary reality, space for instance, and the "unreal" stuff of the virtual world. Think of the cold, hard, here and now—reality—and the nebulous, ephemeral, abstract notion of virtual space. For some people, the idea of interacting with an electronic representation is virtual reality. The clash between virtual and real (in its everyday sense) couldn't be any clearer (Heim, 1999). How could we have both in one, the real and the unreal? We enter a reality paradox when we enter cyberspace (Holeton, 1998; Samoriski, 2002).

We are interested to notice that communicators find that changes in their communication lives in the "real" world are significantly influenced by their behavior in the Internet world. Referring to an electronic discussion (in a group called the "WELL") about online personae, Turkle (1999) wrote: "What most characterized the WELL discussion about online personae was the way many of the participants expressed the belief that life on the WELL introduced them to the many within themselves. One person wrote that through participating in an electronic bulletin board people let the many sides of the self show, 'We start to resemble little corporations, 'Logins R Us,' and like any company, we each have within us the bean-counter, the visionary, the heart-throb, the fundamentalist, and the wild child. Long may they wave.' Other participants responded to this comment with enthusiasm" (p. 79). The point here is that a simple real versus unreal dichotomy fails by oversimplification.

Yet another way to interpret cyberspace is to recall that the mathematician Norbert Wiener first applied the term cybernetics to the notion of systems that control or govern themselves. For instance, the output of the system is governed by information fed back into the system. Wiener (1948) referred to this concept as *feedback*. As you will remember from the discussion of communication models (see Chapter 2), the notion of response is central to the process of human communication. And the computer and the Internet have moved the process of feedback to center stage. Over the Internet, human-to-human and human-to-machine interactions are interactive and tied to the idea of information, or meaning, which feeds back. Of course, the feedback on the Internet differs in some significant ways from feedback in face-to-face encounters. And information and meaning need to be looked at closely because information is different from meaning. Through a close look at the process of communication, we hope you will gain a greater understanding of communication on the Internet.

Theoretical Roots

Although the Internet relies on modern computer technology, we can understand computer-mediated communication through theories of the long-standing tensions in human communication. For instance, a group of communication theorists studying intimate communication in close relationships have elaborated the idea of contradictions and multiple realities in describing interpersonal communication. According to their theory of relational dialectics, communication among intimates is better described by the tensions that simultaneously pull us in opposite directions rather than any scientific laws (Baxter & Montgomery, 1996; Montgomery & Baxter, 1998). *Theory* means the explanation or speculation the investigator seeks to build through investigation. Theory supplies the framework in which we can observe and speculate, ask questions and give answers.

Contradiction is the central concept of the theory of relational dialectics. A pair of close friends may want to spend more time together one moment, for example, and feel the friendship is smothering at another moment. There is a tension between connectedness-separateness, between certainty-uncertainty, and between openness-closedness. Relational communication is better described by multiple realities than a singular, law-like reality. Each individual perceives truth from a personal perspective. Researchers of the relational dialectics perspective find that friends, lovers, and family communicate within systems of contradiction (Rawlins, 1992).

From the perspective of a *linguist* studying discourse, another scholar arrives at the idea that "communication is a continual balancing act, juggling the conflicting needs for intimacy and independence. . . . To survive in the world, we have to act in concert with others, but to survive as ourselves, rather than simply as cogs in a wheel, we have to act alone" (Tannen, 1990, pp. 27–28).

Another way to look at contradiction or paradox within human communication is to examine the tension between the *inner-outer dichotomy,* or the forces inside and outside the individual. One could argue that this tension dates back to early Greek thought. Objections to "new technology" of literacy go back at least to Plato's day, when Plato voiced concerns about writing as "inhuman, pretending to establish outside the mind what in reality can only be in the mind" (cited in Ong, 1982, p. 79).

Perhaps the most influential effect of the Internet is change. We believe a better understanding of the Internet can be gained by relating Internet paradoxes with the paradoxes of dialectical tension (e.g., Baxter & Montgomery, 1996; Pawlowski, 1998; Rawlins, 1992). See if this approach works for you. We can categorize dialectical tension—the natural push and pull we experience in developing interpersonal relationships—as similar to the push and pull we experience in communicating via the Internet (Table 3.1).

Today, you can see a similar force dividing the way scholars approach research in the social sciences. At the risk of over-simplification, there are two primary points of view in this tension regarding whether human communication is essentially rooted in *the mind* with its proclivity towards symbolism and meaning or rooted in *the body* with its proclivity toward observed behavior. We believe both the consequences of communication technology—in this instance, the Internet—and human communication itself are characterized by the tension of contradiction.

Humanist versus Behavioralist Approach

The *humanist* examines how the individual makes sense of the world from personal perspectives and conceptions. The research approaches of *transformational grammar, discourse analysis,* and *ethnomethodology* reject *operationalization* and its associated requirement for all *observable variables.* Additionally, these research perspectives take into account underlying structures of the mind.

The *behavioralist* examines human communication as essentially rooted in observable behavior, as witnessed in defining all variables studied in observable terms or *operationalization.* According to the behavioralist perspective, meanings and observable

TABLE 3.1 *Dialectical Interpretation of Paradoxes of Communication via the Internet (Adapted from "Typology of Dialectical Contradictions" (Pawlowski, 1998).*

	Dialectic of Integration-Separation	*Dialectic of Stability-Change*	*Dialectic of Expression-Privacy*
Internal	Autonomy-Connection	Predictability-Novelty	Openness-Closedness
External	Inclusion-Seclusion	Conventionality-Uniqueness	Revealment-Concealment
Internet	Internet Brings People Together vs. Creates Barriers Between People	Convergence of Media Technology vs. Convergence of Communication Modes	Intellectual Property vs. Public Domain
	Technological Determinism vs. Individuality	Communication between People vs. Communication between Person and Computer	Interpersonal Medium vs. Mass Medium
	Global Village vs. National Isolation	Digital Divide vs. Technical Together	Local and Embodied Community vs. Online Community Free of Time and Space
	Cultural Homogenization vs. Cultural Individuality	Increase Health Due to Information Access vs. Decrease Health Due to Becoming an Inactive Internet Potato	Frees Work Time and Space vs. Intrudes Work into Home Time and Space
	Us Domination vs. Equalization of Power Through Group Retribalization	Increase Leisure Time by Reducing Workload vs. Increase Work Load Because of Higher Expectations	Privacy vs. Loss of Privacy
		Reduce Psychological Well-being vs. Increase Psychological Well-Being	
		Increase Worker Productivity vs. Distract Workers from Their Jobs	
		Work vs. Play	
		Reduce Interpersonal Communication vs. Increase Interpersonal Communication	

responses are one, and hence there is no need to look anywhere except outward, at the observable response. B. F. Skinner (1987) and his followers exemplified this position in their behaviorist psychology. The behaviorist perspective finds its methods of research in attempts to study variables that are physically grounded in the observable world, for instance by studying how frequently one writes to a discussion group or what words they use.

One way to characterize this tension between the humanist and the behavioralist is in terms of defining and locating meaning. *Meaning* is the significance or idea that symbols create in our minds, or the way we attach understanding to symbols. *Symbols*—such as words—are representations that stand for something else. Is meaning to be found in the world or is it a mental construction dependent upon nonphysical variables? Table 3.2 shows two points of view.

The humanist versus behavioralist perspectives often take on a distinction in the method employed to study communication. Typically, a researcher brings a humanist *or* a behavioral methodology.

Generally speaking, the humanist perspective is concerned with how the individual experiences the world, their perceptions, beliefs, attitudes, interpretations, in short, *subjectivity*. The behavioral methodology relies upon observed behavior that is objectively measured or counted. A central difference between the two perspectives is what counts as data. Humanistic data is someone's impressions *as they experience them*. And of course, subjectivity can be studied through objective methods—where a sample of individuals is "observed" for their subjective experience of an event.

We will see this tension played out in discussion of Internet effects in terms of emphasis upon meaning-making and individual interpretation versus observable behaviors. Recall the lament of the programming engineer in Chapter 1, who was torn between the cold logic of the machine and the human needs of other people and her own soul searching. The engineer offers her perceptions. This is not a sample of subjects studied for their behaviors, whether outwardly observable or reported by the subject.

TABLE 3.2 *Theoretical Tensions in the Study of Human Communication*

Humanist	*Behavioral*
Mental	Physical
Virtual reality	External reality
Internal Influences	External Influences
Inside the person	Outside the person
Online	Onground
Theoretical	Observable
Cognitive	Social
Subjective	Objective
Intrapersonal	Interpersonal
Synthetic	Analytic
Interpretative perspective	Functionalist perspective
Reject operationalization	Accept operationalization
Descriptive	Normative
Feminine	Masculine
Immaterial	Material
Qualitative	Quantitative
Inward	Outward
Old-fashioned	Youthful
Contingent	Necessary
Inside the black box	Outside the black box

Perhaps the main reason for calling your attention to this humanist/behaviorist tension in the study of communication is that communication on the Internet further intensifies the tension. We are not the first to notice that the computer as a communication technology raises the question of "where is the soul in the machine?" We think that exploring communication on the Internet is as good a route to finding that soul as any other, and probably better. We propose looking into meaning, interactivity, self-definition, and community as starters.

Meaning

We fully understand that not all scholars would agree with the parallels and distinctions of Table 3.2, but these features suggest the ways we can view the Internet. In its own way, the Internet provides a new space frontier through which to explore the tension between various lines of thought. As you read some of the contradictory research findings about the Internet, you will understand the influence of different points of view of specific researchers. The researcher's methodology and theoretical bent—whether humanist or behavioralist—can determine the focus and the outcome of analysis. By *methodology*, we mean the procedure or process the investigator uses to answer her or his research questions. *Theory* is the more specific idea used to explain phenomena. At the present stage of the development of the Internet, it is unlikely that any one theoretical framework is going to accurately highlight all the key issues. Any one theory with strong claims to answering all our questions about communication on the Internet would be suspect.

In the case of studying human communication on the Internet, we see no reason to limit the methodology. We do believe, however, that some research on the Internet is of higher quality than other research. Regarding all research, we think the most useful approach to understand communication on the Internet lies in the exploring the ways in which individuals attribute meaning to their experience. Generalizations are interesting prompts for discussion and sometimes can help predictions, but the way you as an individual create meaning while communicating on the Internet—that is what matters.

Adoption and Diffusion

A glance at the history of communication technology suggests that new technology is typically not embraced by the public immediately. In most instances it is not clear early on just how a new medium of communication will be used (Baron, 2000, pp. 216–246). The *diffusion of innovations* happens in a particular way, as part of a process of social change (Rogers, 1995). Diffusion of Internet use has been influenced by social, economic, political, educational, and personal factors (Winston, 1995). And whether or not a new medium of communication is worthwhile is always a controversial issue. Henry David Thoreau, contemplating the emerging telegraph system in the first half of the nineteenth century, had this to say: "Our inventions are wont to be pretty toys, which distract our attention from serious things. They are but improved means to an unimproved end. We are in great haste to construct a magnetic telegraph from Maine to Texas; but Maine and Texas, it may be, have nothing important to communicate" (Thoreau, 1854). The same concerns are said today about the Internet.

One fact does seem certain, though: Regardless of how Internet use evolves, people who live in the United States are using the Internet more and more. In a study of American use of the Internet, researchers found that use of the Internet is growing rapidly. Two million more people join the ranks of new Internet users each month. These new users include all demographic groups and all geographic regions (A Nation Online, 2002). The majority of America is online.

Informing and Conversing

In her book, *Alphabet to Email: How Written English Evolved and Where It's Heading,* Naomi Baron (2000) explained a useful distinction between informing and conversing, by reference to communication technology. She drew a two-by-two table of communication technology with regard to the social variables that each type of technology enables. A particular communication technology may permit a one-way transmission of the sender's message to the receiver of the message or the technology may permit a two-way dialogue. A particular communication technology may operate as a point-to-point device, with messages exchanged between two people, or as a broadcast device, with messages sent out to multiple receivers of the message (Table 3.3).

This matrix helps us see how some technology is geared toward informing and other technology is geared toward conversing. By *conversing,* we mean interactive, transactive, cooperative, or conversational communication. At times, communication on the Internet is a combination of two-way communication—including conversing and informing. Sometimes two people converse through point-to-point dialogue or a sender involves multiple recipients, such as in broadcast dialogue. Other times, one person informs another through the Internet, such as by using one-way communication between a person or business and a recipient. You have probably experienced a point-to-point monologue, as with unsolicited e-mail and one-way transmission with multiple recipients, web pages, or broadcast monologue. Some of the time, e-mail is between two people, or two-way. Some of the time e-mail is between a person and a group. Web sites are often between a person and many recipients, like in *narrowcasting*—messages directed to a specific audience—or *broadcasting.*

TABLE 3.3 *Technology Geared Toward Informing vs. Conversing*

	Informing: One-Way Communication	*Conversing:* Two-Way Communication
One Person Receiving the Message	Point-to-Point Monologue	Point-to-Point Dialogue
Multiple People Receiving the Message	Broadcast Monologue	Broadcast Dialogue

Obviously, there are numerous variations or interactions of communication variables on the Internet. The point here is the configuration itself has implications for the nature of the communication event. With the Internet, the balance is dramatically shifted more toward conversing than informing. The Internet is much like interpersonal conversation. Conversing seems an obvious feature of the Internet, but interaction bears powerful influences for communication.

Some scholars of the media see the Internet as particularly inviting to people who had been raised on mainstream media and who had come to distrust mainstream media because of their control and censorship of information, and for the portrayal of what they consider negative values, ideology, and imagery. Fidler (1997) speculates "an underlying distrust of mainstream news media and other traditional sources of information appears to be one of the contributing factors in the growing popularity of the Internet and other consumer online networks" (p. 112). In comparison to the mainstream media, Internet users could play a meaningful role in creating their own realities online, in "speaking" rather than just "listening," and in finding information they consider "the truth."

Advocates of the skeptical view of the Internet see the open communication of the Internet as an era that has already ended. Now, the skeptics maintain, information is increasingly protected and controlled. According to the cynics, new hierarchies of power are forming in the Internet. Critics distrust the commercialization of the Internet. Do you think they are right?

Mowlana (1995) writes forcefully about the tension between the Internet as a means of empowering individuals and the Internet as a means of developing a centralized control of information. The fear is that the Internet is displacing indigenous cultures with a Western, secular, and consumerist culture (Arquette, 2002). Many people do not have access to computers or computer networks: "We must make sure that the promise of digital technology is not lost to a world in which basic telephone service is still too expensive for most people" (Mowlana, p. 6). Consider the effects of the corporatization of information.

We cannot ignore the political, cultural and economic forces that operate with the communication forces on the Internet. As you read our text, we encourage you to keep in mind Mowlana's claim that "Globalization is but another word for the impending triumph of American culture: entertainment, fashion, and the American way of life, all combined in one package" (p.1). A basic issue in all of globalization is to consider how people will use the Internet in the future.

Reflection

What are some theoretical roots that may be helpful to the study of human communication on the Internet? Although these roots provide nourishment to our understanding, they are just a vehicle for carrying the nourishment. We need more ideas, more interactivity, more analysis to feed the theoretical roots. What else do you need to know to adequately examine human communication on the Internet?

Case for Discussion

Janet's 27-year-old sister was accidently killed in a car wreck. Thieves accessed information about Janet's sister online, and quickly created fake identification, fake credit cards, and fake financial accounts. The thieves stole money from Janet's sister by stealing her identity, then quickly accumulated bills the financial institutions attributed to Janet's sister. Janet found herself in the process of fighting financial institutions and creditors as she tried to clear her dead sister's name. "The pain of losing my sister is horrible, but the agony of having her identity stolen and used after her death makes everything so much worse."

Richard Power (2000) discussed how the Internet provides extraordinary potential for common criminals, terrorists, and nation-states. He talked of a single crime in Russia that stole $10 million from a bank. Power told of the inability of government and business to provide the necessary countermeasures to effectively defend themselves. Of course, business and government sources are concerned about terrorism via the Internet (Sager et al., 2001). With hundreds of millions of web pages on the Internet and the First Amendment's protection of freedom of speech, there is little that can be done to protect people. Or is there?

Internet Investigation

1. *Interplay of Paradoxes.* By *interplay,* we mean a going back and forth, sometimes in a light or playful sense. Do you consider the Internet fun, for example? Hard work, irritating, controversial? Involving or boring? Stimulating or mind-numbing? An asset or an interference to your quality of life? Discuss some of the paradoxes interplaying on the Internet.
2. *Cultural Influences.* In this chapter, we discussed how globalization is the process whereby people of the world become increasingly interdependent. Investigate the relationship between the Internet and globalization. Post your findings to your online discussion group.
3. *Dialectical Tensions.* To investigate the concept of dialectical tension, first consider the term *dialectical,* which refers to talk—the discourse, dialogue, or interactive communication between people. *Tension* is the conflict or paradox that naturally occurs in that communication. Consider these examples. We want to feel close to another person, but we don't want to give up our individuality. We want to be open in what we say, but we don't want to give up our privacy. We want to be able to predict the other person's behavior, but we enjoy novelty and surprise. Investigate research in dialectical tension and give examples of how communication on the Internet fuels similar tensions.

Concepts for Analysis

We have provided concepts relevant to this chapter so that you can check your understanding. These concepts are worth dissecting in more detail. As a research option, examine one concept. Use our citations as a starting point for your study. We recommend that you search for refereed articles in an online database. Find at least three solid references you can read. Summarize, scrutinize, and present your analysis to other people in your course.

Questions to stimulate your inquiry: Do you agree with the ideas as we presented them? What other points of view should be considered? What details need to be added to fully understand the concept? What can you contribute through your personal analysis of the concept? What more do scholars need to know about the topic?

A *behavioralist* is one who examines human communication as essentially rooted in observable behavior, as witnessed in defining all variables studied in observable terms or operationalization. According to the behavioralist perspective, meanings and observable responses are one, and hence there is no need to look anywhere except outward, at the observable response (Skinner, 1987).

Broadcasting is the process of sending a message to a large group of people through the medium of television or radio. With the concept of DirectTV, cablecasting innovation, and computer convergence, the concept of broadcasting is changing.

Cyberspace must be a metaphor for some sort of imagined environment or psychological experience.

Diffusion is the process of social change involved in the adoption of an idea, practice, or object. In this case, we are concerned by the diffusion of Internet use, which has occurred at a rapid rate in the United States (Rogers, 1995).

Discourse analysis is used broadly in this book to refer to a variety of schools of thought that take a close look at utterances produced in natural, spontaneous conversation, with an interest in understanding both structure and function of utterances in everyday conversation.

Ethnomethodology refers to research methods aimed at discovering what people think and feel about particular communication events as they occur in ongoing social contexts.

Feedback is the communication response. As a public speaker, you'll want to observe audience reaction, facial expressions, head nodding or shaking, restlessness, and any cues that give you information about their response to your message.

Globalization is the process whereby nations become increasingly dependent regarding military intervention, finances, and communication.

A *humanist* perspective examines how the individual makes sense of the world from personal perspectives and conceptions. The research approaches of transformational grammar, discourse analysis, and ethnomethodology reject operationalization and its associated requirement for all observable variables. Additionally, these research perspectives take into account underlying structures of the mind.

Inner-outer dichotomy refers to the forces inside and outside an individual's mind: Is human communication essentially rooted in the mind with its proclivity towards symbolism and meaning or rooted in the body with its proclivity toward observed behavior?

Linguistics is the study of language.

Meaning is the significance or idea that symbols create in our minds, or the way we attach understanding to symbols.

Methodology is the procedure or process the investigator uses to answer her or his research questions.

Narrowcasting is the act of directing messages to a specific audience, such as an Internet site with a narrow audience.

Observable variables refers to variables—things that can change—that can be measured, as opposed to forces that are believed to operate but cannot be measured.

Onground means face-to-face in a physical place, not online.

Operationalization refers to the procedure—what is actually done—for the measurement or manipulation of a variable.

Subjectivity is the idea that communication and assignment of meaning is personal, variable, individualistic, and mental.

Symbols—such as words—are representations that stand for something else.

A *theory* is the explanation or speculation the investigator seeks to build through investigation. Theory supplies the framework in which we can observe and speculate, ask questions and give answers.

Transformational Grammar is a theory of grammar that takes into account both surface and underlying structure, i.e., observable and unobservable variables.

Part II

Functions

In this section, we discuss basic functions of communication via the Internet. We see three major functions: conversing, informing, and playing. As part of its paradoxical nature, we explore how the Internet can polarize people.

    Chapter 4—Informatics—Gives you ideas about how the Internet is used as a communication vehicle for information. We discuss research and scholarly communication, including the value of critical reading and peer-reviewed journals. You will learn about the importance of using quality primary sources. And you'll learn how to find those sources through databases and search engines.

    Chapter 5—The Play of Internet Communication—Gives insight into the ways people play online. We discuss computer games, metaphors, and other types of interplay between people and communication principles.

    Chapter 6—The Polarization of People—Gives you a perspective of much of the dark side of communication. The Internet can cause a digital divide, dissonance, and violence. People can use the Internet to violate your privacy, spread rumors and hoaxes, steal, and more. We talk about the nature of flaming online and suggest reasons why we think flaming occurs. We also explore the use of the Internet by hatemongers and terrorists.

4

Informatics

If you were going to open a new restaurant selling healthy fast foods, would you do research before selecting a location for the restaurant?

If you were going to offer an athlete a spot on your team, would you do some research into her or his athletic ability first?

If a journalist wrote a story, would you expect her or him to research the situation?

Of course you would, because information is essential to valid understanding and quality decision making. In this chapter, you will examine how people can communicate through the Internet for the purpose of obtaining valid information. The Internet provides a nearly limitless source of information, enabling access to communication never before available to the average person.

Focus

```
TO: Students

FROM: paradoxdoc@hotmail.com

RE: Overview: Informatics

In chapter 3, Lenny and Joan discussed the Internet
functions of conversing and informing. Although their
```

perspective is a bias toward conversing, now they discuss the role of informing.

Informatics is the ability to communicate information through technology, a way of managing information through computer systems. In this chapter, Lenny and Joan discuss informatics as it relates to investigation, examination, analysis, search, and observation. They discuss the potential of the Internet to offer high quality information, including peer-reviewed scholarship, library resources, and converging video, audio, and textual offerings, which are available via your computer, while providing suggestions for increasing your digital literacy.

Lenny and Joan discuss refereed journals, e-journals, and the peer review process. They suggest ways to find quality sources, which have validity and reliability.

Lenny and Joan suggest you learn how to use a variety of databases, which can help you find high quality information online. A database refers to a collection of information, in this case, electronic journals, typically organized by subject field and journal types. Full-text databases are particularly useful because they make the full text of periodical content available. Most professional databases are funded by subscription, and often we encounter an online database through the online library at schools, colleges, businesses, organizations, and local libraries. In contrast, an Internet search engine is a way to find information on the Internet (comparable to a card catalog), while a database is the body of information available (comparable to the reference works, journals, and books in a library). Generally, databases have consistently higher quality information than the Internet (where anyone can provide a Web site).

Lenny and Joan offer suggestions for increasing your information search skills. They talk about their favorite online databases, such as Expanded Academic and MasterFile Elite. In addition, Lenny and Joan give ideas about how to use the Internet for scholarly research.

In this chapter, you will explore how people communicate scholarly information through the Internet. This communication advances knowledge because the Internet intensifies research capabilities and information exchange. This information should help you to do online scholarly research. At the end of this chapter, you should be able to answer the following questions.

1. What is informatics?
2. What is the value of research communication through the Internet?
3. How can the Internet be used for gathering scholarly communication?
4. What are quality, primary communication sources?
5. How do scholars communicate information through databases?
6. How can you effectively search for information online?
7. What information is communicated through journalism and Internet scholarship?
8. How might a person communicate scholarly research through the Internet?
9. How do people communicate through scholarly discussion groups?
10. What paradoxes are evident in informatics?

What Is Informatics?

Informatics is the ability to communicate information through technology, a way of managing information through computer systems. Informatics uses digital data storage and retrieval. We can go a step further, in that informatics is the transformation of data to decision making. When we examine the Internet, that transformation from information to decision is a complicated one, with many paradoxes involved. Thus, the study of informatics includes the analysis of the nature, structure, and support of information on the Internet as a foundation for how people know, attach meaning, and live their lives. *IT* is an acronym for "information technology" or the use of electronic means to send and receive information. IT is an all-inclusive term to mean computers, computer systems, and the Internet. "IT people" are what you may call "techies," or the people who work on the technological side of computer systems. Communication scholars, in contrast, are concerned about informatics from the standpoint of the quality of information they receive.

The Internet has made information acquisition much easier. It intensifies an individual's research capabilities by enabling users to obtain information and to advance knowledge. With the Internet, information can flow in both directions: (a) researching the Internet to obtain information, or (b) providing information to others through the Internet.

Research is a word that covers a broad expanse of ideas. Research is investigation, examination, analysis, search, and observation. People use the term to mean looking into just about anything by just about anyone, in just about any way. You might check out a neighborhood you are thinking of moving into; you might talk to a broker about a stock you are thinking of buying; you might look into a prospective employer's likes and dislikes, or you might read *Consumer Reports* to find out about the merits and demerits of a car you are considering purchasing.

Used in these ways, *research* is roughly synonymous with *finding out about something*. And finding out about something is essentially what scholarly or academic people do when they conduct research. Scholarly uses of the term "research," however, adds a considerable number of ifs, ands, and buts to the everyday use of the term. *Scholarly research* is a systematic investigation that employs rigor, research methods, replication of findings, statistical analysis, review of the literature, collection of data, consistency of observations, accuracy, testing, predicting, defining terms, cause-and-effect, observing and analyzing, theoretical explanation, and critical review. This distinction between research as "looking into" in the everyday sense and research as scholarship will play a critical role in this chapter.

Research Communication

The most popular ways people communicate through the Internet are for e-mail, entertainment, and information (Petersen, 1999). A perusal of books about the Internet and how to use it shows that accessing information on the Web largely—though not entirely—refers to non-scholarly sources (Ackermann & Hartman, 1999; Ashton, Barksdale, Rutter, and Stephens, 1995; Barron & Ivers, 1996; Doyle & Gotthoffer, 2000; Estabrook, 1997; Flanders & Willis, 1998; Gilster, 1994; Harris, 1994; Lambert & Howe, 1993; Stanek, 1998). And so we may fail to see that tucked away in some corner of this virtual world of casual and not so casual communication is a realm of scholarly communication. And we may fail to realize that certain populations of Internet users truly value scholarly communication online. College students, for instance, are increasingly required to take courses that make use of online resources, such as course web sites, with a syllabus, schedule, links to readings, and so on. In one study, college students ranked learning as their most valued benefit from their use of the Internet (Perse & Ferguson, 2000).

You probably already have done some looking into something on the Internet, such as finding out what time a movie is playing, or how much a book may cost, but what about scholarly research? Have you conducted the kind of research you would expect to find in a journal article or described in a college textbook? Have you used the Internet to read scholarly articles? How much scholarly research is available online? What is the potential of the Internet for storing scholarly research and for offering it up to anyone interested in reading rigorous inquiry? Notice we are not talking only about research about Internet communication, but about research available through the Internet. Because our focus is upon scholarly research and the Internet, we find that the need for critical reading of Internet information is increasingly important, especially as online journalism grows (Kopper, Kolthoff, & Czepek, 2000). The importance increases because through the Internet persuasive messages reach us, education develops, and access to information of all kinds increases. What is the potential of the Internet for students and teachers to find and use scholarly articles? What is the potential of the Internet for scholars to add scholarly research to the Internet? This chapter will explore just these sorts of questions. From our communication perspective, informatics is receiver-emphasized communication. By that we mean that the responsibility for effective communication lies with the receiver. The receiver has to find the information and evaluate the validity of the information (Radford,

Barnes, & Barr, 2002). We should remember that much of the impetus behind the initiation of the "Information Superhighway" legislation, beginning in 1988, was to enhance education and research (The National Research and Education Network, NREN). Consider this quote from an update of this plan.

> It is clear that the education and library communities are continuing to expand their use of electronic networks. Researchers, students, librarians, and educators subscribe to electronic conferences, newsletters, and journals on a wide range of topics of concern to them in their work. They use electronic mail to communicate with remote colleagues, file transfer to acquire a variety of public domain information resources, such as software and full-text files; and remote login to access supercomputers. In addition, members of the library and education communities are contributing to the creation of electronic information resources and services. For example, many library catalogs and databases created originally for local use are currently available over the Internet. Further, some librarians and educators are an important source of network training and support for their clientele. (Bishop, 1991, para. 6)

Accordingly, imagine scholarly communication on the Internet as roughly divided into three functions:

1. Gathering scholarly information,
2. Contributing scholarly information, and
3. Communicating and collaborating.

Scholarly Communication

Most people seldom hear about the rich scholarly information resources available on the Internet. Instead, they hear people speaking about the dangers of the Internet, and how students need to be protected as they cross the minefield of ideas. Perhaps you are concerned with bad people who are trying to hurt innocent folks or destructive ideas on the Internet that may be placed there for the public to view. All of us have been warned of the need for protecting our privacy online and for reading critically because anyone can publish her or his ideas online. While there is basis for our fears, we can confuse a medium with its content if we see the Internet as the problem. We can use the Internet to enhance our work performance, maintain effective relationships, and improve our quality of life by providing access to needed information. The medium holds the potential to offer high-quality information, including peer-reviewed scholarship, library resources, and converging video, audio, and textual offerings, which are available via your computer (Jones, 1999).

Too often sites designed to help the user assess the quality of an online document bring attention to the superficial trappings of the document, when it was last updated, who the author is, and whether she or he has a Ph.D. The quality of the author and publisher is not the substance of our analysis of research quality. The quality of the article's information itself is the substance of our analysis of research quality. Writing about web pages that carry biased and hateful views, such as anti-Semitic, racist, or homophobic sites, Gurak (2001) reminds us that "Many of these sites masquerade as informational, adopting the format and style of a newspaper, report, or other truly informational setting. They use professional-looking fonts, high-quality layouts, and other credibility-boosting devices that are easy to employ on the Web. But to be cyberliterate means to be able to look past the format" (p. 56). There is, however, excellent help for students who seek to develop systematic approaches to their online information research (e.g., Basch, 1998; Ray, 1999). Critical reading of scholarly information online requires the same ability used in critically reading paper journal articles. Before we proceed further, we need to review some terms.

Refereed Journals

Refereed journals are scholarly publications that are usually the most reliable source of information. The articles considered for publication typically are primary sources because they are written by the scholars who conducted the original research projects. The journal editor then sends the articles to experts for review. If the reviewers believe the research information and ideas are solid, they will approve the article for publication. This refereed information can be published in e-journals or paper journals made available online through a database.

Critical Reading and Peer Review

The argument that students must read critically what they receive online should be seen as simply a reminder that all scholars must read critically. Again, the book as a form does not guarantee accurate information, nor does the journal article. Whether the text is online or on paper does not change that fact. As for the argument that paper scholarly journals follow a peer review process, assuring some degree of quality control, the process of peer review of scholarly online journals is no different. In fact, online journals that do institute peer review are in a position to carry out the process electronically in most instances, moving manuscripts to reviewers online, receiving reviews online, and notifying authors online. In fact, it has been argued that online publication of scholarship is subject to greater feedback from readers because the publication is highly accessible with interactive capabilities. Readers can e-mail the author and the author can continually update, adjust, and advance the article on the web site. Ultimately, the same principles of good research scholarship apply online as on paper. It may even turn out that students with access to online scholarship improve their critical thinking skills.

Once we move past what appears to be anxiety over dealing with a new medium, the central question in gathering scholarly information comes down to simply this: "How do I find the scholarship?" While this chapter is not intended as a manual, we will offer examples of some excellent sites, less with the intention of serving as a reference manual and more with the intention of pointing out the rich scholarly resources that are now available online. Of course, these sites can and likely will change and disappear with time. But once you know that the scholarship is there on the Internet and how you can benefit from it, you can easily find useful new sites.

Quality Primary Sources about Communication

Quality primary sources contain well-constructed arguments. Some rely upon evidence provided by others, their studies and writings, and some report on evidence gathered by the author. Solid sources are valid, relevant, concrete, and ethical. As a rule, the closer you are to the original source, the more reliable the information, although there is no substitute for your own critical thinking about the quality of the article. The original source is considered a *primary source,* meaning it is written by the person who conducted the research being cited, or the individual who experienced the event being discussed. Primary sources are the most useable information sources for the scholar, since they typically explain how the evidence was gathered. Generally, primary sources are more valid than secondary sources, since there are fewer intermediaries involved. Primary sources provide the information professionals in a field need to know so that they can replicate the research. A primary source of information is the foundation of information that represents the original report of the observations or research (Frey, Botan, Friedman, & Kreps, 1992).

When you read any research, you will want to consider the quality of the research design and procedures. E-mail *survey techniques* have generally been less successful, for example, than postal mail survey techniques. When reading research using survey techniques, you need to pay attention to how the survey was conducted and the number and sampling procedure of respondents. If a researcher simply puts a survey online at a web site, for example, the sample of people who go to that site might be quite skewed or biased in some way. If the researcher uses prenotices, questionnaires, thank you/reminders, and replacement questionnaire techniques, then the response rate is probably as good for e-mail surveys as postal mail surveys (Schaefer & Dillman, 1998). But this is only one example of what to look for in reading research done using the survey technique. Other procedures to look at carefully include: the way the questions are asked, the order, the choices given, the wording, the person doing the questioning, the context, and many others.

The quantity of information available through the Internet is enormous, but remember, you want *quality sources.* First, don't feel overwhelmed. Second, use helpful information locators. Third, be sure to evaluate the *validity* of the sources. You can find plenty of incorrect information in books and other hardcopy sources, but on the Internet, anyone can create or post to a site with no editorial review and the content can be totally fallacious or false. As you read on, we will help you to seek quality information. How will you find *reliable* sources of information? Reliability is consistency. You'll need to evaluate the quality of resources you use to make sure that the information you report is consistent with other reliable sources of information. Are they accurate, reliable, up-to-date? There are some web sites designed to help you evaluate sources.

Online Databases

Thirty years ago, academic scholars had to learn elaborate systems of academic communication. Researchers would use reference books to find articles to read. Then they would seek the articles in their own library. If the article wasn't available, they would have to travel to a library that contained the resources they needed. Today, through the Internet, a scholar can accomplish in a matter of hours what used to take days or weeks of effort. Databases permit quality searching to find information, while many provide the full text of articles the scholar needs to read.

An online e-journal refers to a journal, usually much like a paper journal, except that it is accessible via the Internet. Many—but not all—scholarly journals are peer reviewed, and many require a subscription. A database refers to a collection of information, in this case, electronic journals, typically organized by subject field and journal types. Most professional databases are funded by subscription, and often we encounter an online database through the online library at schools, colleges, businesses, organizations, and local libraries. Note a search engine and a database are two different entities. A search engine is a way to find information on the Internet (comparable to a card catalog), while a database is the body of information available (comparable to the reference works, journals, and books in a library).

Many students find that learning how to use online databases is one of the most valuable skills they obtain in college. Through databases you can find the answers to questions you have about business, entertainment, education, even your personal relationships. In essence, online databases enable you to access all kinds of information from the comfort of your own home, provided you have Internet and database access. We believe that effective scholarly communication may be one of the most valuable aspects of human communication on the Internet. So, we present this information as an integral part of the foundation for this book. The databases are invaluable. Not only can you access this information efficiently from home or work, but you have the advantage of being able to print, save to disk, or e-mail the articles to yourself, friend, colleague, or co-worker. We believe that learning what databases are available and how to use these databases effectively may be one of the most important communication skills you learn in this course. This chapter will take time and effort, but you can walk away with a valuable life-skill: *The skill to find any information you need!*

Most colleges and many organizations and local libraries pay subscriptions to access Internet databases. Although the exact content of each database varies widely, you can find information in the sciences, arts, politics, and any other topic you need. For databases that contain *popular magazines, trade magazines,* and *scholarly periodicals,* the hardcopy of the periodical is electronically scanned so that it can be provided online. Sometimes you will find odd errors because of the *scanning* process. One student, for example, put a quote in his essay that read "There is a dear link between . . ." The quote didn't make sense, but if you give a little thought to the statement, it's easy to recognize that the computer scanner simply read the word clear as dear. Thus, you'll need to watch for scanner errors.

There are many databases available, and although we don't pretend to have covered them all, we have included the major ones you are likely to need. A computerized *database* is a collection of information—citations, definitions, abstracts, full-text articles, data—from magazines, journals, books, and other reference information. Typically, databases are organized by subject, author, title. Because they are computerized, however, databases can search using a variety of additional criteria, such as year, word in the article, journal the article was published in, peer-reviewed journals only, and those kinds of details. *Refereed journals* are scholarly publications that are usually the most reliable source of information. The articles considered for publication in refereed journals typically are primary sources because the articles are written by the scholars who conducted the original research projects. Refereed journals are available as paper and online information sources. Some databases, for example, are loaded onto local library computers or available on CD. Our emphasis here is on databases available through the Internet.

Typically, professional databases are funded by subscription, but you may be able to access an array of online databases through the online library at schools, colleges, businesses, organizations, and local libraries. Remember, a search engine and a database are two different entities. Increasingly, college students use computerized versions of academic journals because their college or university provides access to online databases of abstracts, full-text journals, and reference materials. For newcomers to this technology, the comparison between going into their library to find a journal article and bringing an article home with them versus finding a journal article online is an eye-opener. For instance, at a university where winter hits pretty hard, just getting to the library in wintertime can be an undertaking, where icy conditions and parking problems are only a start. Actually finding the article, viewing it on a microfiche machine, and copying the article are often a challenge. In contrast, searching an electronic database takes merely a few keystrokes. And often, receiving the full text at home is simply a matter of e-mailing the article to yourself.

Full-Text Databases

A number of large companies offer databases that make full-text journal articles available on the Internet. In comparison to paper sources, there are advantages to online databases. Internet databases have:

1. Easier access to full-text articles.
2. More complete searching abilities.
3. Better access to large collections of information.
4. Accessibility from distant locations.

Your library may have more than a hundred online scholarly databases. Table 4.1 is a list of our favorites, although there are many other excellent databases. Because Internet communication is an interdisciplinary field in the social sciences, you will find relevant information in the areas of communication, ethnicity, sociology, psychology, information technology, and anthropology.

All of these database companies have subscription fees, but your local library, college, business, or other organizations may have accounts that will give you access to such databases.

Skill Section: Online Search

Although each database may be slightly different in what it contains and how it operates, there is much commonality between databases. You can follow these steps to conduct an effective search.

TABLE 4.1 *Popular Databases to Find Full-Text, Peer-Reviewed Information*

Expanded Academic Index

Expanded Academic Index http://www.galegroup.com/tlist/sb5019.html contains more than 500,000 citations to articles, news reports, editorials, biographies, short stories, poetry, and reviews appearing in some 1,500 periodicals. Subjects covered include the humanities, social sciences, general sciences, art, cultural studies, economics, education, ethnic studies, government, history, literature, politics, popular science, psychology, religion, sociology, and literature relevant to African, Asian, European, Latin American, and Middle Eastern Studies. The SearchBank version contains some full-text articles.

GenderWatch

Where it says "Record Type Phrase," select "Journal." Http://gw.softlineweb.com/ (GW) is a full-text collection of approximately 140 publications and archival material including international journals, magazines, newsletters, regional publications, special reports, and conference proceedings devoted to women's and gender issues. GW provides in-depth coverage of subjects that are uniquely central to women's daily lives, such as family, childbirth, childcare, sexual harassment, aging, body image, eating disorders, and social roles. It also includes content on the impact of gender and gender roles on areas such as the arts, popular culture and media, business and work, crime, education, research and scholarship, family, health care, politics, religion, and sports.

MasterFILE Elite

Go to http://search.epnet.com/login.asp and select "Expert Search," then scroll down to select limit to "Full Text" and "Peer Reviewed." MasterFILE Elite provides abstracts and indexing for more than 3,100 periodicals, plus searchable full text for more than 1,200 periodicals. Subjects covered include general reference, business, health, and more.

Social Sciences Full Text

http://www.unesco.org/general/eng/infoserv/doc/shsdc/journals/shsjournals.html Social Science Full Text provides indexing and abstracts for articles of at least one column in length from English-language periodicals published in the United States and elsewhere plus the full text of selected periodicals, Abstracting begins with periodicals published in January 1994. Full-text coverage begins in January 1995. Topics covered include addiction studies, anthropology, area studies, community health and medical care, corrections, criminal justice and criminology, economics, environmental studies, ethics, family studies, gender studies, geography, gerontology, international relations, law, minority studies, planning and public administration, policy sciences, political science, psychiatry, psychology, public welfare, social work, sociology, and urban studies.

1. *Select a database* that is relevant to the kind of information you seek.
2. Depending on the database and what you seek, you may want *to narrow the hits* to a manageable number (e. g., 50–100). Most will give you information beginning with the most recent. If so, you'll be able to scan the beginning of a long list to see if your search is on track. You enter key words for your search. Often

there are options you can select, such as the date, journal, peer-reviewed, and full text. You also may be able to select what the database searches: author, title, abstract, or full text.

A combined or *Boolean search* will enable you to find a manageable number of abstracts or articles that are on target. A Boolean search enables the combination of terms by using these connectors:

AND NOT
OR
AND
WITH
ADJ (for adjacent terms)
NEAR (for terms close in order)
SAME (for words in the same paragraph)

3. *Examine the hits* you have for clues to what is working and what has failed. If you are having trouble with finding the right information, be sure you are using the right words through the database's thesaurus. Many fields use jargon, which may be quite different from common terms. If you look for an article on stage fright in communication journals, for example, you probably will find little information. The term used in this field is communication apprehension. So, make sure you are using the correct terms for your search. Some databases will look at your term, then give suggestions for the key words that might work in that database. You may need to experiment a bit. When you find an article you like, look at the key words used in that article to give you ideas about what you might use in the next phase of your search.
4. When you find an abstract or article you like, you can *mark the items that you might want to use.* You can examine the list of marked items to make sure you're on track. Then send the list to yourself via e-mail (or print or save to disk). If you are using a full-text database, you'll want to send the entire article or simply an abstract. But be sure to read the article abstract to make sure it's really what you want. Otherwise, you may have to wade through a bunch of articles you don't want to read. The *abstract* is a summary of the article, which will give you basic information about the contents of the article. Most scholars don't approve of quoting from an abstract because the abstract may have been written by someone other than the author, so it's not a primary source, and may have misinformation. In addition, the abstract is probably so brief that it has insufficient information to be useful in your research. You will want to find full-text articles for your research.
5. After you have the articles you need, *read* them. You may be seeking the information for your own edification. If you are preparing a business report, academic paper, or a speech, for example, you can cut and paste information from the article to your paper. The way you do this is to open your word processor and cut and paste the information. *PDF* files are photographs and cannot be copied in the same way as regular text files. If you are working on a computer away from home, you can save or e-mail the material to use later.

6. Be sure to *use quotation marks and cite complete reference information exactly.* If you fail to cite the source of information you quote or paraphrase, that is *plagiarism.* There are many online sites that will give you information about how to avoid plagiarism.[4]
7. *These listed steps assume that you find the exact information you need the first try, which is unlikely.* Thus, you may need to narrow or refine your search. Some ways to focus include:

- Examine the key words used in an article that is what you want, and use those key words to search.
- Remember, many databases have a thesaurus of specific terms they use, which can help you find the right academic jargon to unlock the information you seek.
- Select a database with focused content.
- Use a search engine that is more focused.
- Select only English language articles (or whatever languages you speak).
- Narrow to recent or specific dates.
- Select only abstracts or full-text articles.
- Search by author, title, or subject.
- Use the author names, references sited, or keywords from one article that provides useful information.
- Conduct a Boolean search that combines multiple words connected.

E-Journals

We have already noted that an online *e-journal* refers to a journal, usually much like a paper journal, except that it is accessible via the Internet. Individuals who cannot access a scholarly database must seek reliable online information from free e-journals, library membership, or subscription. We have found that more and more e-journals require a subscription, but there still are many e-journals that are free to the public (e.g., Online Electronic Journals and The Internet Public Library). The Electronic Journal Miner, in fact, specifically offers the user the option to ask for free online journals. Although full-text refereed articles may be harder to find, it is still common for the public to have free access to abstracts of journal articles and citations to sources.

There are sites where you may find some online journals providing full text without a fee or other resources for communication and the media. But a search of the e-journals from most sites will most often show restricted use. At the same time, there are many resources at the sites that are useful and open to the public.

So, you should assume that you will find access to excellent databases. You may be surprised by the sophistication of some small libraries, which gain database access through cooperative library, school system, and government consortiums. Many colleges give alumni access to their databases. In some cases, residents of certain counties—whether college students or not—have full access to major university database subscriptions.

Database use is an area definitely worth investigating now and in the future, no matter where you live. In our experiences, there is no one more helpful to your educational and professional career than a talented reference librarian! Find a reference librarian whom you can count on and seek help through the Internet maze.

Journalism and Internet Scholarship

Like so many other aspects of online scholarship, the newspaper online is evolving. Research about online journalism is in its infancy (Kopper, Kolthoff, & Czepek, 2000). Currently, one can access hundreds of newspapers online, in a wide range of languages and emanating from all over the world. The online reader can explore many sites.

> Associated Press at: http://wire.ap.org/public_pages/WireWelcome.pcgi
> The Chronicle of Higher Education at: http://www.chronicle.com/index.htm
> CNN.com at: http://www.cnn.com
> Online Newspapers around the globe at: http://www.onlinenewspapers.com
> The Washington Post at: http://www.washingtonpost.com

Many more online newspapers and databases could be listed here. ProQuest, for example, provides full-text articles from an array of newspapers. You may find valuable information through Ethnic NewsWatch, which provides the full text of many ethnic newspapers in the United States. The Internet has become a means of communication for journalism, both to acquire information in writing stories and to disseminate news stories. But it is difficult for researchers to investigate what is going on because technology and user patterns change so quickly. Very often the available research is more relevant to market questions than online journalism per se. And much of the research information is privately funded and not publicly available.

As for the writing itself—the content and structure—there is a scarcity of studies to guide us. This scarcity is true not only of journalism sites, but about scholarly communication in general on the web. You can find writing on how to persuade or sell, how to create graphic design, how to structure web pages for navigation and hypertext architecture (Kilian, 1997; Sammons, 1999), or how to please web site readers (Perse & Ferguson, 2000), but relatively little on quality of content online.

> We may try to impose some kind of hierarchical, linear order on it, but the online medium remains essentially hypertextual. Just as the basic unit of hypertext discourse is the "chunk," a screenful of text or graphics, the basic unit of online learning is some kind of freestanding idea, almost like a sound bite. It may link to other related concepts, but arbitrarily. The student may choose some personal, even perverse sequence of learning. We can't do much about it unless we undercut the whole principle of hypertext and parcel out our chunks of knowledge in a calculated, teacher-controlled sequence. (Killian, 1997)

Again, journalism is little different from other areas of Internet information research, at this point, because journalism relies more on case studies than scientific methods.

Another area of online journalism and scholarship that will evolve in the future is *interactivity*. The ability of readers to interact with the medium and with other readers is a big difference from traditional media. Readers can talk to one another and to journalists/scholars with various online applications accessible on web sites. Potentially, people can use e-mail, discussion boards, or chat rooms. Now journalists can poll readers on various issues and report the results online. Again, development of these technologies in Internet journalism is in process, and we cannot predict the end result. As Callahan (1999) suggested, the Internet is the most important reporting tool since the telephone.

We can see that textbooks are offering online sites where students can find a host of information. For instance, a popular textbook for the introductory course in communication is Em Griffin's book, *A First Look at Communication Theory*. On the book's web site the reader finds a user survey; information about a video displaying interviews with the theorists discussed in the book (the video is also on CD that comes with the book); movie clips; primary sources; a "Meet Em" web site; e-mail links to Em as well as to others closely tied to the book, and more. It is even possible to access full-text chapters that appeared in earlier editions of the book but have since been edited out. *Archives* by theory topic are at: http://www.afirstlook.com/bigdisp2.cfm?pass1=arch.htm&pass=archther.cfm

Some textbooks provide the instructor with slide presentations to accompany the chapters, making the teacher's preparation for courses much easier. Allyn and Bacon/Longman, for example, offers an array of online materials, a course environment through Course Compass, online tests, and PowerPoint slides for courses. Imagine a course where you study the textbook, view the video on their computer, review particular chapters of the book, and then go to the course's online discussion to talk about questions posed by the instructor or just to collaborate with one another, your instructor, and the authors about course ideas. Allyn and Bacon/Longman has a public speaking course e-package—Public Speaking Unbound—that offers these elements in interactive text modules. This process may be quite similar to how you are learning in this course.

Communicating Scholarly Research Online

If gathering scholarly research brings us to databases of e-journals and academic listservs, then contributing scholarly research must mean that we need to place scholarship in these same places. Because a database is a collection of journal articles or other online resources, contributing communication means submitting manuscripts. We think it is important that you understand the submission process, so that there is no mystery to all of this—just how information becomes accessible to you. Your instructor may have contributed information to such sources, for example. Interested scholars can find out how to do that by visiting e-journals and reading the submission policy, just as they might when looking into submitting to a paper journal. For instance, journals will provide information on how to submit a manuscript, such as the information offered at the online site of the juried (or peer reviewed) journal, Currents in Electronic Literacy.

Currents in Electronic Literacy encourages submissions that take advantage of the hypertext and multimedia possibilities afforded by our World Wide Web publication format, as well articles concerning the use of emergent electronic technologies. To this end, we gladly accept articles with graphics, sound, and hyperlinks submitted as HTML documents. We ask, however, that such submissions adequately consider reader-access issues. For instance, we ask that submissions incorporate such accommodations as the inclusion of <alt> tags in an image and the use of content tags (e.g., citation <cite> and emphasis <em> tags) instead of the corresponding physical markup tags (e.g., italics <i> and bold <b> tags) whenever possible. For detailed information about making Web documents accessible to people with disabilities, please refer to the World Wide Web Consortium's Web Content Accessibility Guidelines 1.0 at http://www.w3.org/TR/WAI-WEBCONTENT/.

Currents is also pleased to publish essays in more traditional formats. Please submit these either in HTML format, Word 97/98 or 2000 format, or Rich Text Format (RTF). We accept electronic submissions by e-mail at ejournal@lists.cwrl.utexas.edu as well as on 3.5″ floppy or Zip disks by post sent to the following address:

Currents in Electronic Literacy
c/o Computer Writing and Research Lab
Parlin 3, University of Texas at Austin
Austin, TX 78712

Again, there is no difference in the process of peer review when the journal is online versus on paper, or there need not be any difference in the quality of information. If you examine the quality of the scholars on the journal editorial board, that should give you insight into the quality of the journal's peer-review process.

There are other ways that a scholarly manuscript might appear online. Some journals are not peer reviewed. We have been involved in editing an online publication of the National Communication Association, for example, *Communication Teacher Resources Online*.[5] In this instance and many other sites like it, the editor may decide what to include in the periodical or a staff of editorial assistants may take part. The reader may feel that there is some degree of control in such cases because—depending upon the periodical and who is in charge—there is a centralized system for allowing in some items and keeping others out. And yet another way in which articles appear online is by individuals uploading their own work to a server and making it available. This technique is the way that we hear about most when people say "anyone can put anything on the Internet." Perhaps this process is akin to a vanity press, where the author can publish her or his ideas without going through the peer-review process.

Online Scholarly Discussion Groups

There are many opportunities to share and test information and ideas online. For student interaction online, we find that discussion lists are more appropriate than chat. The lists lend themselves better to more serious and on-task discussion (Honeycutt, 2001). Academic groups may very well deserve to be understood as a different creature from other groups. There are different ways in which we can think about online discussion groups: *list activity, academic versus commercial, climate, and leadership* (Rafaeli &

Sudweeks,1996). Scholarly discussion groups are often used to articulate ideas, to find others who will agree with one's ideas, to convey information, to test ideas, and to hear others' ideas. *Scholarly discussion groups* are a sort of hybrid of a seminar, library, conversation, and newsletter. College students seem to enjoy synchronous and asynchronous discussion (Davidson-Shivers & Tanner, 2000). Kelly, Duran, and Zolten (2001) found that reticent students compared to nonreticent students preferred e-discussion to classroom discussion. But if you have participated in this kind of scholarly discussion, you already understand some of the aspects we have discussed. You may have your own Web site and discussion group. You may have conducted extensive scholarly research and published online.

Reflection

How do you use the Internet to receive information? Do you approach research and scholarship in an organized, methodical, and systematic way? Do you understand some of the ways information is communicated on the Internet? One of the most important skills you can learn in college is how to access and evaluate information through the Internet, so you can find the answers to your own questions. As you learn effective use of informatics, you will be able to answer your questions;

- When you need to know how to approach a problem at work . . .
- When you need ideas about how to increase your commitment in a personal relationship . . .
- When you want to find out anything . . .

TABLE 4.2 *Paradoxes of Informatics*

High-Quality Sources	*Low-Quality Sources*
Refereed journals.	People accept invalid Internet information without questioning.
Quality information is more accessible.	Only people with certain technology and database access can acquire the information they need.
Highly credible information.	Anyone can make invalid content appear credible.
Search engines like Google.com can enable you to access an array of information quickly.	Many search engines give results based on revenues paid to the engine, so their communication is profit motivated.

You can use the Internet!

There are paradoxes in this communication for information and scholarship. You can find excellent scholarship online, but you can also find scams, false information, and other communications that lack validity (Table 4.2).

Despite the problems with the validity and credibility of much Internet information, you can learn to be a discerning communicator. You can use search engines like Google.com to find web pages. The search engine you choose can be an important decision because many search engines are fueled by client payment not by free access to all content (Introna & Nissenbaum, 2000). What are the politics of information (e.g., Dutton, 1999)?

Databases tend to be less political and more reliable than searching the open Web. Databases will allow you to read books, articles, and abstracts from popular, scholarly magazines, and other sources. The Internet medium holds the potential to offer high quality scholarship, peer reviewed scholarship, library resources, and converging video, audio and text. You can potentially access books, magazines, newspapers, atlases, photographic files, and anything else you can access in a library.

Ultimately, the same principles of good research scholarship apply online as on paper. Solid sources are reliable, valid, relevant, concrete, and ethical. Usually, the closer you are to the original source, the more reliable the information. The original source is a primary source, meaning it is written by the person who conducted the research being cited.

How many databases have you actually used? In the next ten years, what kinds of questions will arise in your life that you can answer through research in these databases? We hope you have learned the importance of databases, discovered the accessibility of information in full-text databases, and want to explore an array of databases and URLs. We think the best way to learn about these databases is to use them. Your college library may have access to a large variety of databases, which you can continue to access as a college alumnus. Your local public library may be part of a consortium that enables access to amazing resources.

If we sound enthusiastic about the value of these resources, it's because we are! Most people fall into habits and patterns of online research that they teach themselves. By taking the time to learn what databases are available and how to use them effectively, a world of information opens to you—literally. Take time to explore and develop your research skills, which can be among the most useful skills you learn.

Case for Discussion

In your course discussion group, give your analysis and perceptions on the following.

Dan was not particularly excited about taking a certain communication course. He'd heard there was much research and writing involved. The first week of the course the class met in a computer lab on campus. The course professor and a college librarian showed the class how to evaluate Internet sources, the pros and cons of valid sources, and what databases were available. By the end of the week he said to the professor: "This is the most practical course I've ever taken. Now, I can find out the answer to anything."

a. What are the values of being able to find the answers to your own questions?
b. How will you learn the resources available on your campus?
c. What standards will you use to judge the validity of Internet sources?

Internet Investigation

1. *Scholarship Online.* Conduct online research about scholarship on the Internet, particularly finding valid sources on the Internet. Prepare an essay or discussion group posting on the topic. Cut and paste in the URL of the source of information you used.
2. *Evaluation Criteria.* Write a list of criteria you will use to evaluate the validity of online scholarship during your research. Then look at this site. Feline Reactions to Bearded Men http://www.improbable.com/airchives/classical/cat/cat.html This site looks like solid research. What is your analysis?
3. *Database Exploration.* Go to Course Compass and access each of the databases listed in Figure 4.1. Explore three databases in significant detail. Write a brief essay explaining what you found and post the essay to your course discussion board.
4. *Search Engines.* Learning search engines is a good place to start learning your way around the Internet. Try each of these search engines. Select the three you like best and the three you like least. Explain why in a class discussion.

AllTheWeb: alltheweb.com
AltaVista: www.altavista.com
Argus Clearinghouse: www.clearinghouse.net
Direct Hit: www.directhit.com
Excite: www.excite.com
Google: www.google.com
Guidebeam: guidebeam.com
HotBot: hotbot.lycos.com
InfoMine: www.infomine.com
Invisible Web: www.invisibleweb.com
IxQuick Metasearch: www.ixquick.com
Librarians' Index to the Internet: lii.org
Northern Light: www.northernlight.com
ProFusion: www.profusion.com
Query Server: www.queryserver.com
Surf Wax: www.surfwax.com
Vivisimo: vivisimo.com

5. *Mass Communication View of Politics and the Internet.* The access and nature of information and communication can be quite political. Mass communication theorists are concerned with the politics of the Internet. What is the relationship between information and politics? When information and ideas are transmitted, does the Internet create more or less politicalization of ideas? What is the political effect of the Internet?
6. *Discussion.* In your course discussion group, answer the following questions.
   a. How can you effectively use online databases?
   b. How do you access, search, and find peer-reviewed full-text information?
   c. What are some databases you used after reading this chapter that you have never used before?
   d. What are some tips you can give others in effectively using full-text databases?

Concepts for Analysis

We have provided concepts relevant to this chapter so that you can check your understanding. These concepts are worth dissecting in more detail. As a research option, examine one concept. Use our citations as a starting point for your study. We recommend that you search for refereed articles in an online database. Find at least three solid references you can read. Summarize, scrutinize, and present your analysis to other people in your course.

Questions to stimulate your inquiry: Do you agree with the ideas as we presented them? What other points of view should be considered? What details need to be added to fully understand the concept? What can you contribute through your personal analysis of the concept? What more do scholars need to know about the topic?

Asymmetric information is a theory that suggests that seldom do both sides have the same information.

An *abstract* is a summary of an article that gives you basic information about the contents of the article.

Archives are online collections of newspapers, speeches, or some other data.

A *Boolean search* combines two or more terms when conducting a search of an online database or search engine. Terms are most typically combined with "and" and "or" and similar connectors.

A *Database* is a collection of information, often abstracts, full-text information from journals, books, and other reference information.

An *E-discussion list* is a computerized format used by a group of people to converse online. Often the term is used interchangeably with e-group, meaning any group of people who talk online. Their purpose may be personal, business, or educational. These online groups or lists may communicate synchronously or asynchronously via electronic mail, course environments, listserves, or webboards.

The term *E-journal* refers to a journal, usually much like a paper journal, except that it is accessible via the Internet.

A *full-text* article is a complete article, which is available through an online source. Excellent full-text databases include Expanded Academic Index, GenderWatch, MasterFILE Elite, and Social Sciences Full Text.

A *hit* is an appropriate source found through an online search.

Informatics is the ability to obtain information through technology systems.

Interactivity is the ability of readers to interact with the medium and with other readers.

IT is information technology or the use of electronic means to send and receive information. IT is an all-inclusive term to mean computers, computer systems, and the Internet.

Online means using a computer that is connected to another or many computers through an intranet or Internet.

On paper—in contrast to online—means hardcopy or information that is in some form of paper format.

PDF files are photographs and cannot be copied in the same way as regular text files.

Plagiarism is the uncredited use of information quoted or paraphrased.

Popular magazines are ones designed for a general audience and with less rigorous investigative techniques.

A *primary source* is one written by the person who conducted the research being cited or who experienced the event being discussed. Primary sources of information represent the original reports of the observations or research.

Quality sources are reliable, valid, relevant, concrete, and ethical.

Refereed journals are scholarly publications that are usually the most reliable source of information. The articles considered for publication typically are primary sources.

Reliability is consistency. Quality sources consistently give the same information over time, information consistent with other reliable sources of information.

Research is a word that covers a broad expanse of ideas about investigation. Research is the process of studying, investigating, examining, and scrutinizing the world in an effort to advance knowledge.

Scanning is an electronic copying process used to duplicate material. For example, for online full-text articles, the journal articles are scanned electronically and made available in a computerized format.

Scholarly discussion groups are a sort of hybrid of a seminar, library, conversation, and newsletter.

Scholarly periodicals are monthly, quarterly, or annual publications of articles written using scientific methods of investigation.

Scholarly research is a systematic investigation that employs rigor, research methods, replication of findings, statistical analysis, review of the literature, collection of data, consistency of observations, accuracy, testing, predicting, defining terms, cause and effect, theoretical explanation, and critical review.

A *search engine* is a way to find information on the Internet (comparable to a card catalog).

A *secondary source* is one provided by someone other than the original author or researcher. Secondary sources are considered more likely to be flawed than primary sources.

The *survey technique* is a method of data collection that asks questions of people using questionnaire or interview techniques.

Trade magazines tend to be quality periodicals designed for professionals in a certain occupation.

Validity is the accurate measure of what it is supposed to measure.

Notes

4. *Plagiarism: What It Is and How to Recognize and Avoid It.* Available at: http://www.indiana.edu/~wts/wts/plagiarism.html

5. http://www.natcom.org/ctronline/index.htm

5

The Play of Internet Communication

A teenager in an arcade battles pterodactyls in a multi-user virtual reality. Two pensioners on different continents exchange e-mails and images of their grandchildren. A heavy-metal enthusiast in the United Kingdom orders a compact disc over the Net from a retailer in the United States. And a class of 11-year-olds research an article from a dozen worldwide web sites and publish the result on their school's intranet.

But stop! Wait! Help! Somebody is enjoying themselves! People are communicating who may never have met! Companies are selling directly to consumers at low prices! And young people are learning to interact with knowledge and to think for themselves rather than to passively accept linear, single media programming!

—(Barnatt, 1999, para. 1–2)

These are the kinds of users who experience *intensified interplay* while communicating on the Internet. Do you enjoy playing video games? Do you have fun shopping online? Do you fantasize about what someone looks like when you meet them online? The operative words here are enjoy, fun, and fantasize. The Internet can magnify, elaborate, and enhance certain elements of human communication and play is one of those elements. The entertainment value of the Internet naturally creates the play of online communication (Dickinson, 2003). And although we can call the Internet an entertainment medium, it is much more, including a source of news, information, instruction, and persuasion. That sense of play, however, can carry into all functions of the Internet.

The way we think is interwoven with the way we communicate. Pesce (2000) wrote: "When we change the way we see and touch the world, we change ourselves" (p. 248). The

Internet influences not just outcomes but the process of communicating and by doing so it influences how we think. Just as interactive toys engage a child, interactivity on the Internet increases the sense of "someone" listening to us, observing us, reacting to us personally, and thereby encourages us to work and play harder (Pesce, 2000). Turkle (1984) too argued that computers engage our minds as playmates and intellectual interlocutors. More and more theorists are suggesting that with the Web, we are living in our heads. The new technology, the Web being high on that list, facilitates our imaginative and playful thinking. Writing about the computer and simulation in his book, *The Playful World: How Technology Is Transforming Our Imagination,* Pesce points to the increased opportunity to pretend. Pesce (2000) wrote:

> We are at the threshold of a revolution in human experience. Computer simulation is starting to be used as an engine of the imagination, bringing to light some of the most intangible aspects of our being. Just as music, dance, theater, and poetry have helped us articulate the quiet parts of ourselves, simulation will become a new aesthetic, an art form with its own power to illuminate the depths of our being. (p. 248)

We must wonder if simulating face-to-face communication over the Internet doesn't "articulate the quiet parts of our being" and spark and transform our imagination.

Focus

```
TO: Students

FROM: paradoxdoc@hotmail.com

RE: Overview: Play

Lenny and Joan discuss the positive and pleasurable
feelings prompted by many types of communication on the
Internet. They use the term "interplay" to mean an
interaction, which can be transformative, entertaining,
interesting, or light-hearted. Multi-user dimensions,
such as MUDs and MOOs, are one type of play on the
Internet. But Lenny and Joan have observed several types
of Internet play: Imaginative play like a child, identity
role-playing, interplay with the self or other, play like
```

> theater, games for destruction, playful work, and the fun of something new. Lenny and Joan elaborate in a discussion of the play of computer games.
>
> Lenny and Joan also discuss the many types of metaphors in online communication. They talk about consistent stories, myths, and virtual interactions of e-discussion groups. IT—an abbreviation for Information Technology—also has sexual connotations. In addition to sex metaphors, they give examples of Internet language metaphors that suggest animals, bodily functions, and violence.
>
> Lenny and Joan explain their understanding of the interplay between the Internet and communication principles. Internet communication is a process. Adapting to the Internet can increase communication effectiveness. Other concepts addressed include: Communication is irreversible. Internet meanings are in people. No one can experience totally effective communication on the Internet. Finally, in this chapter's skill section, Lenny and Joan offer tips on how to design a web site.

In this chapter, we will focus on how the Internet can intensify communication play. At the end of this chapter, you should be able to answer the following questions.

1. What is the play of Internet interplay?
2. What are associations of playing computer games?
3. How does the Internet intensify the use of play through metaphors?
4. How does the Internet intensify fantasy?
5. What is the interplay between human communication principles and the Internet?

The Play of the Internet

Do you like exploring the Internet? Is online research drudgery for you? Does the Internet energize you? Lull you to sleep? Give you pleasure? Or seem more like a pain in the neck? Does time pass quickly while you interact on the Internet? If you have experienced some of these contradictory responses, then you experience the paradoxes of Internet communication (Table 5.1).

Millions of people have access to the Internet through home, work, or school, or they can use a local library to access the Internet, where they can access a free e-mail account, moderate a free e-mail list, maintain a free web page account, and learn via online tutorials

TABLE 5.1 *Which Terms Best Describe Your Particular Use of the Internet?*

fast	slow
easy	hard
complex	simple
exciting	boring
suits my self	suits others
interesting	uninteresting
mental	physical
many sensory	single sensory
fun	work
imaginative	dull
intelligent	unintelligent
freeing	confining
social	asocial
multiple stimuli	single stimuli

(McDermott, 2001). Of course, not everyone has automatic access to the Internet in this country, and certainly around the globe most people do not have access to the Internet. In most cases, the Internet requires English literacy and Internet literacy. And there are people who avoid using the Internet, including some people who use the Internet only because they are required by a specific employment or educational situation. Our theory of Intensified Interplay probably cannot be applied to these people, at least not at this stage of their Internet use. The kind of Internet user we are discussing in this book—the kind of person who intensifies interplay through the Internet—we speculate, would tend to select descriptors from the left column of Table 5.1. The concern here is to analyze the communication processes employed by people who are regular, interested users.

Do you have positive and pleasurable feelings or negative and unpleasurable feelings about the Internet? Internet play is connected to symbolism. *Symbolism* is the representation of meaning through words, actions, or other means. We use the term *interplay* to mean an interaction, which can be transformative, entertaining, interesting, or light-hearted. We select this term partly because the word interplay is often used in communication studies to represent communication interaction. We also use the word interplay because of the potential sense of play on the Internet. We are reminded of the work of Stephenson (1988) and his theory that people attach meaning to the world based on pleasure or unpleasure (displeasure). The idea is that any words, ideas, or communication are symbolic, and the way we attach meaning is through how we feel about them. Stephenson referred to people using computers as *computer pscience*. The p is silent just like the p in the word psychology. The p recognizes the way computer use creates internal or intrapersonal processing in addition to the actual computer processing (Stephenson, personal conversation). To extend his theory, we suggest that even when using the Internet for work and information purposes, people respond to the Internet according to their feelings of pleasure. Like other mass media, the Internet is an entertainment and information medium, but the Internet creates an interplay between people. Thus, the play of interplay is intensified by the Internet.

Although offering a significant theory to analyze mass communication, Stephenson was not the first or only person to recognize the importance of play in communication. More recently, for example, Schroeder (1996) contended that play operates within its own space. He agreed with Huizinga's theory of play (1950) and contended there are four elements to the structure of play:

1. Play is for itself. It serves no external goal.
2. Play exists outside the scope of ordinary life.
3. Play operates within fixed boundaries of time and space, with its own set of rules.
4. Play is pliable. Though it can completely absorb the player, "ordinary life" can reassert itself at any time. (Schroeder, 1996, para. 18)

We have *rewritten* these postulates for the Internet.

1. Internet play is for the self. It serves internal goals.
2. Internet play co-exists inside and outside the scope of ordinary life.
3. Internet play operates without fixed boundaries of time and space, although the play may operate within Internet rules.
4. Internet play is pliable. The Internet can completely absorb the player so as to integrate with "ordinary life" or take on a life of its own.

Turkle (1995), who was centrally concerned with how we build identity on the Internet, writes extensively about various aspects of play, since, she points out, play is integral to the construction of identity. We play with identity on the Internet. Turkle is especially attentive to play in MUDs, although she does not limit Internet play to MUDs. We select a character for ourselves, a context in which to act out our character; we choose a gender, occupation, history, age, and communication style. We are what we say. We are authors in a story we tell along with others.

Special attention should be given to the ways in which play and games, as in role playing games, may have powerful effects on who we are and who we become as a consequence of play. Keep this is mind as you read on. As you read, try to imagine ways in which play can have real effects on people in their real lives, online and off.

If you agree with the idea that play is highly relevant to communication on the Internet, then you may wonder: How does play *function* on the Internet? We have identified seven types of play during human communication on the Internet:

Imaginative play like a child.

Identity role-playing.

Interplay with the self or other.

Play like actors act in a play.

Games for destruction.

Playful work.

The fun of something new.

Imaginative Play Like a Child

Circulating humor, jokes, having fun, and playing games are some of the kinds of Internet play available to users. Some people respond playfully to creating virtual experiences online, perhaps because they already use and enjoy the medium for computer games, MOOs, and MUDs. Human-computer interaction (HCI) often involves play when individuals interact with the technical equipment as well as people when they communicate via the Internet. When individuals name their computers, for example, they are interacting with the technology. People can interact with computers in ways that are similar to the way two people communicate (Reeves & Nass, 1996). Perhaps you have played a game of chess with a computer, for example. You experienced human-computer interaction when you played this game. Perhaps you interacted with a robot—also called a "bot"—while you were in a MOO. MOOs and MUDs are multi-user dimensions. People interact with each other as if they are together in a real space. For example, a group of online friends may imagine they are together in a fraternity house, or a group of students may imagine they are together in a real classroom. They have to move through space in the place. Participants may take on certain characters or identities for their interaction. In addition, there may be computer "robots" in the dimension. MUDs tend to be a playful interaction—like a game—and these robots will be programmed computer questions, characters, or responses to the real people who are interacting in the mud. The people who engage in these MUDs can become extremely involved in their interactions with each other (N. M. Kendall, 1998). MOO and MUD Internet user groups exist for educational or entertainment purposes and often develop elaborate rules of interaction. Frequently the users employ significant fantasy in the interaction, even pretending to be together in a real place or pretending to be certain characters. This kind of e-discussion has been called a cyberspace playground because the interaction is like child's play (Scodari, 1998). Some Internet users have playful entertainment as a goal in the Internet communication. Yet, there is "a playful aspect to discussion groups" and chat (Barnes, 2002, p. 41).

There are web sites designed for playful purposes too. By way of example, see Fun with Grapes,[6] a site that pokes fun at academe or the related Strawberry Pop-Tart Blow-Torches[7] site. Consider for a moment the child-like play suggested by the following web site.

> Jokes aside, after its first three months MeowMail[8] boasts 10,000 registered members. Cat owners can go to the site to send and receive e-mail from their cats and download cat visuals, such as a fish tank screensaver. Each week, The Morning Hairball has features including an advice column in which well-wired cats (via their owners, or human valets) can ask when it's appropriate to throw up or what to do when tempted to eat your owner's hamster. (Kornblum, 2000)

Identity Role-Playing

The Internet allows users to experiment with their identity. If you talk online anonymously, if you pretend to have a different job than you really do, then you are role-playing through the Internet. The role-playing may be a part or extension of the self.

How does online discussion promote role-play? Some Internet users experiment with gender switching, pretending to be someone else, sexual fantasy, and other types of sexual interplay. Some people on the Internet lie, experiment with their identity, pretend to be people they are not, even try to speak from the point of view of the opposite gender (Turkle, 1997). We have been surprised, when talking to some e-group members, to learn how easily and often some people pretend online, as if that is simply part of the game of the WWW.

Interplay with the Self or Other

Scholars often refer to human communication as interaction, intercourse, or interplay (e.g., Adler, Rosenfeld, & Proctor, 2001; McIntosh, 1995). You sometimes hear people who use interplay to mean communication. We hope to arouse certain connotations with the use of the word interplay. The words interaction, intercourse, and interplay bring slightly different ideas to mind than does the word communication. We intentionally use the word interplay because of the root play. We believe that Internet users *approach their communication activity* with a sense of playfulness, and users seek ways to use the Internet that are pleasurable.

Television soap operas (daytime serials), for example, are continuous stories that are quite involving for some viewers. Those stories relate to Internet interaction because many soap operas have cyberfan pages and discussion groups (Baym, 2000). These discussion groups give viewers a forum for talking about the plots, characters, interactions, and events. These e-group discussions give the story lines and characters a kind of validity or reality as the interactants talk about them almost as if the characters are real people. Viewers also discuss the actors and their personal lives. Some actors find the discussions upsetting and disturbing because of their critical and inflammatory nature. While most cyberfan groups value free speech and freedom from control, some scholars are concerned about that freedom's ability to silence certain people through the tendency to convey dominant values at the expense of older women and minorities (Scodari, 1998). Do we have free speech or forbidden speech?

Play Like Theater

Virtual reality is the creation of an imaginary space in which people can pretend. Certain MOOs, MUDs, and games allow an individual to take on a new persona, become a character, be someone else, or transform into someone imaginary. Of course, one can question: "To what extent are the play and actors real or imaginary?" In these multi-dimension user groups, the people involved perform much like actors or players in a scene they might perform in a theater.

Games for Destruction

There are people who play games on the Internet which are designed to destroy the lives or resources of others. The spectrum ranges from gossip to terrorism. Although the destructive effects vary, the gossip, virus transmissions, online stalking, S & M sex sites, and violent videogames often take on a sense of gaming for the Internet users involved, so that they receive the adrenaline rush of winning a game that takes something away from other people (time, status, money, feelings of safety, employment, life, power, innocence, control).

Playful Work

How is the Internet an interplay of play and work? Modern college faculty know that if divisions between learning and play were ever clear to students, that clarity no longer exists. Faculty use popular films to demonstrate communication concepts, computer simulations to demonstrate points, online groups to discuss theoretical principles. And any faculty member who has taught in a computer lab has noticed students taking diversions to check fun web sites and e-mail from friends during class. This digression is probably no more prevalent than ever before. One classic study of students listening to faculty lectures, for example, found that at any given moment, a large number of students were daydreaming or fantasizing about something else (cited in Lucas, 1993). Today students' mental flights can travel through the Internet. And when that same medium is used for completing challenging research and learning, the student may create an interplay of pleasure and work which is intensified by the Internet.

The Fun of Something New

One man told us that he was a "gadget person" and really enjoys computers. In a survey we conducted, we asked people to indicate feelings they often felt while using their computers. The feelings of satisfaction, fun, productivity, and frustration were the most frequently named (in that order). So long as pleasant feelings occur more than negative ones, users probably feel that positive feelings outweigh the negative ones. In fact, the frustration may be what makes users feel so satisfied. When they figure out the problem on their computer, they feel complete, confident, and satisfied with themselves. More than one respondent to our survey noted a feeling of challenge during computer use (Table 5.2).

You might wonder, "With all this talk of play, experimenting with identity and role playing, how do we distinguish between play and flat out *lying*?" If I tell you that I am a stockbroker and I am not, is that play or lying? If I tell you that I am a woman and I am a

---

TABLE 5.2 *The Principles of the Internet's Intensified Interplay*

1. Internet interplay intensifies cognitive aspects of human communication, specifically, cognitive collaboration.
2. Internet interplay intensifies social paradoxes of human communication.
   a. For some people the Internet intensifies their connection to others, while for other people, the Internet polarizes people.
   b. For some people the Internet intensifies personal involvement through advocacy, while for other people, the Internet isolates people.
   c. For some people the Internet intensifies risk and fear, while for other people, the Internet is a way of coping with threats and reducing fears.

man, is that play or lying? If it is lying, isn't that unethical behavior, far removed from the connotations of play? These are questions that ought to be taken seriously, for lying is a serious matter in the world of human communication. Let's begin by considering how we might distinguish between the two behaviors, lying and playing.

Making the distinction between lying and playing affords us a wonderful opportunity to argue once again that the Internet is a highly intensified medium for cognitive behavior. Think of lying as expressing ideas to another person by communicating not only propositions or claims about the world but additionally, communicating that the speaker believes in the truth of what she or he is saying. That is, if I tell you that it is snowing out, and I say that with the intention of your believing that I believe that it is snowing out, and I *don't really believe* it is snowing out, then I am *lying*. I am not mistaken, because I don't really believe it is snowing out. I am not playing because I want you to think that I believe in what I am saying, and I want you to believe it too—that it is snowing out. Now, if it is in the middle of July when I say this and it is 98 degrees out and you know where I am located, I am either a very bad liar or more likely, I don't expect you to believe that I believe it is snowing out. I must be playing, and I want you to figure that out (Grice, 1975).

If you follow this logic, then you see once more that expressing meaning and assigning meaning is an event that requires our minds to work, whether consciously or otherwise. On the Internet, with the reduced cues of the physical environment, reduced nonverbal cues of facial expression and so on, and contact with people who we may not know so well, it becomes an effortful cognitive event to figure out if the other person is lying or playing. Once again, we see how intrapersonal communication—assigning meaning—and computer-mediated communication are closely related. We might add that another paradox presents itself here: Play on the one hand and unethical behavior of lying on the other create yet another tension in communication on the Internet. It may be that the best way to explore these ideas is to find instances in Internet-mediated communication where it is difficult to decide which way to think about a communication event: as a lie or as play. If you encounter actual instances, these would make for exciting discussion. Have you ever lied or played on the Internet? Consider these two examples.

One interviewee told us that she always pretended to be something other than what she was in reality whenever she entered a chat room or discussion group for enjoyment. The woman told us about an array of identities she used. In one case, she told people that she was a U.S. scientist at NASA. In fact, she was an international graduate student in the social sciences in the United States. When we asked her if this behavior was lying, she replied: "No, it's all just a game. Someone would have to be stupid to think that I was a NASA scientist. I don't speak English that well, and I don't know anything about science. It's all a game." The intention was experimentation and fun, so this example seems to be play.

One interviewee told us that he met a woman from Russia online. She was a physician, and he brought her to the United States so they could marry. It turned out she was not a physician, and had no intention of marrying him, and she cleaned out his bank accounts before disappearing somewhere in the United States. The intention was deceit, so this example was clearly lying.

Play of Computer Games

What is the effect of playing computer games? Clearly, many people in our society enjoy spending time playing games on the Internet. Young men playing videogames, for example, is one image of Internet play that comes to mind. Americans spend $400 million annually on interactive entertainment games such as Microsoft's 2001 Xbox video-game console, which converges DVD, computer, and videogame elements (Elkin, 2001). Although these are primarily self-contained units, video and other types of games on the Internet also are popular.

Although one would expect much creative play when using video games, they may be among the most confining types of play on the Internet.

> Video games provide none of the open-endedness of regular play. In most video games, that programming determines violence as the primary problem solving option, the only way to get points. (Schroeder, 1996, para. 6)

Much of the research about video games has been about potential negative effects of the violence. This body of research is contradictory, inconclusive, and influenced by various theoretical positions.

> Video games have no benefits, but are not responsible for significant problems, either.
> Video games are a neutral technology, capable of good or bad, depending on the actual programming and usage.
> Video games are beneficial in an information age, teaching new literacies and skills and providing auxiliary physical benefits. (Schroeder, 1996, para. 11)

The different points of view may be a function of the researchers' biases and methodological points of view. The same range of reactions has been said about the Internet. But our concern here is less about the outcome of play and more about the process of play. We focus on the process so each individual can shape experiences in the way they choose to enable whatever outcome the individual desires.

Although significant in terms of time and money, online video games are not the only types of games you might play on the Internet. Casino games, fantasy football, trivia games, solitaire, and others are among the formalized games on the Internet. These games are the paper-and-pencil or board games of today. One young child, for example, excitedly told us that she learned at preschool that she could play the game hangman on a piece of paper. The child drew a picture of the gallows and letter spaces on a piece of paper as she explained: "I thought that only worked on the computer!"

Playful Metaphors

Much of the communication online is simply for entertainment. We have found that some lists have consistent stories, myths, and virtual interactions. For example, on one list, when participants wanted to imply something "naughty," they'd say they were going "under the chart." On another list, there was a virtual band with several members playing different instruments. None of this interaction was face-to-face, but all in their imaginations.

Those who study communication have long believed that our language affects the way we understand our environment (Carroll, 1956; Sapir, 1921; Whorf, 1941, 1956; Zhifang, 2002). If we examine the language of the Internet, we can find many examples of metaphorical language which affect the way people communicate (Stefik, 1996). The Information Superhighway—or I-Way—is a metaphor for the Internet, which conjures up images of travel with other people to access information. The I-Way seems to be an incomplete metaphor, which has been expanded to include The Digital Library Metaphor (The I-Way as Publishing and Community Memory), The Electronic Mail Metaphor (The I-Way as a Communication Medium), The Electronic Marketplace Metaphor (Selling Goods and Services on the I-Way), and The Digital Worlds Metaphor (The I-Way as a Gateway to Experience) (Stefik, 1996). These metaphors represent different functions of communication on the Internet. In a sense, they offer a contextual way of looking at the Internet.

How does the Internet intensify play through metaphors? In a sense, the Internet is a youthful and immature medium. It's not yet a teenager in terms of development. We are not surprised to find certain metaphoric language and online behaviors that reflect the spontaneity and silliness of youth. A *metaphoric word* is one that draws an analogy or a comparison. The word Web, for example, brings the image of a spider on a web. More than one Internet user has called herself Spiderwoman or himself Spiderman. Whether or not an individual is aware, Internet metaphors create perceptions. Although there has been some examination of the evolving language of the Internet (e.g., Greenhalgh, 2000), the research seems scarce. Much of the analysis of language on the Internet has involved specialized jargon, translation between languages, and programming language (e.g., Dyrli, 2000; Kushner, 2000; "The power," 2000). Cherny (1996) examined the use of language in MUDs—multi-user dimensions—as a way of creating power online. But what about studies of metaphoric language? More than a dozen years ago, Veryard (1987) suggested a link between sex and language that operates in information technology, which is often called *IT*.

> "IT" is not only the neuter pronoun, but is also used to denote sex appeal. Hi-tech is supposed to be "sexy." Once, the word "computer," like the word "typewriter," referred to the man or woman operating the machine; now it refers to the machine itself. According to the metaphors discussed above, a magnetic disk or database contains data and is therefore female; a program manipulates data and is therefore male. The traditional diagrams of these objects correspond to this: disks and databases are shown as rounded cylinders, while programs are shown as hard-cornered boxes with arrows for the input and output.

An interesting note about the metaphoric language of the Internet is the fact that the French government has created a list of terms to substitute for English computer terms (Medintz, 2000). These word alternatives express different metaphors: A startup e-company is a "young sprout," and the Internet is "the spiderweb."

We examined Internet sites and computer programming textbooks looking for metaphoric language. Our analysis of Internet metaphors suggests several categories. This list does not portend to represent every type of computer or Internet metaphor, but represents a sampling of the array of terminology. Below are examples of each category.

Animal: larval stage (early stage of computer obsession), like kicking dead whales (difficult computer work), mouse droppings (computer information given out, also called a cookie).

Bodily functions: bag on the side (refers to colostomy bag, a negative term about functionality), barf (operation failure).

Violence: list-bomb (enrolling a person in a discussion list without permission), locked and loaded (disk in the drive and ready to go), bomb (quit), brute force (primitive programming style), bobbit (place holder), and terms using dead, such as deadly embrace (standstill), and dead link (nonoperational).

Sex: fuck me harder, FMH (misoperating software), SEX (Software EXchange).[9]

We have to be careful about generalizations because of the exceptions they ignore, but we can imagine an adolescent boy chuckling over the use of these Internet words. In fact, these kinds of metaphors seem consistent with the same age, gender, and behavior common to many video game players. Although we can make a case for the intentionality of the violent nature of metaphoric language on the Internet, we suspect that these Internet metaphors often are used in ways that is intended to be humorous and playful.

Example Metaphor

To gain a sense of what it is like to be part of a playful online group, consider these participation rules of an actual online group (Figure 5.1). We have changed names for confidentiality. Note that the technique of substituting words for real words is called metonymy (Agre, 1994).

This information about behavior and language use is sent to the new subscriber after she or he begins receiving e-mail. You can see how these are rules of the online discussion game to be played. The wingflap argot and not being allowed to say the word "relationship" can be confusing and make the new subscriber feel like an outsider. The moderator's intention in avoiding the word "relationship" is to protect those who have been hurt by their breakups.

Internet interplay intensifies imagination and creativity. Fantasy and play are intensified through three elements of the Internet: stories, interaction, and language. Bormann (1980) suggested that by sharing stories, individuals create a group fantasy that describes their group culture. These *fantasy themes* tell much about the people involved in communication. In fact, the fantasies of the story telling are probably more meaningful and important than the reality of what actually happened. The Internet gives us global patterns and communication themes. The Internet is the ultimate group fantasy: The Virtual World.

One *synchronous interaction* example typifies the sense of playful bonding that can happen online through virtual reality. In this case, a person on the list just broke up with her lover. When the conversation started, there was that sense of a coffee klatch as members chatted—perhaps because the people were actually drinking coffee while talking online at their computers that Saturday morning. Before long, a dozen people joined in the conversation, all e-mailing synchronously as they imagined themselves sitting in the per-

FIGURE 5.1 *Participation Rules*

With over 300 BIRDS in the roost, we've found that to keep things safe and to keep feathers unruffled, it helps to have a few simple guidelines. There are also some conventions peculiar to BIRDS, which we'd like to share with you. So:
Nest-iquette—A Guide for Navigating the Roost

WORDS AND PHRASES

Early days in the forest we grew weary of the overuse of the word "relationship," a subject of common constant concern to so many older birds, and so we began to substitute other descriptors in place of the dreaded R-word.
These we put in brackets <> (not parentheses). For instance, you might say:

my <teapot> always says. . .
my <heaven on earth> and I went to. . .
our <toaster oven> was wonderful until. . .
my ex-<flame thrower> taught me. . .

"Wingflaps" are what BIRDS send one another for support, like the equivalent of an e-mail hug. BIRDS request them for all sorts of reasons, and BIRDS are not hesitant to send them out to one another. We are a supportive cyber community!

"Core brag" is what one posts to share a triumph or something wonderful happening in one's life. We don't hesitate to come to the community for support in times of distress, but sometimes we need encouragement to share the good things as well. Thus the concept of "core brag" was born to give us all permission to say them out loud.

We encourage you to use your weekly core brag allowance, whether you think you need to or not. Even if it's just a time when something went OK, when things didn't fall down on your head, we hope you'll jump off your branch for a minute or two and tell us the good stuff too.

BIRDS come into the forest from every conceivable background and some are very sensitive to swear words. For them, we use the convention of replacing letters with asterisks. Sh*t! P*ss! D*mn! H*ll! Well, you get the idea.

Some abbreviations are commonly used over the net:

LOL....Laughing Out Loud
ROFLMAO....Rolling On the Floor Laughing My A** Off
BTW................By The Way
YMMV.............Your Mileage May Vary
IMHO...............In My Humble Opinion
IMNSHO.........In My Not So Humble Opinion

Various emoticons replace the facial expressions we can't see, helping to make the tone of our posts even more clear:

<g>...............grin
<f>................frown
<BG>............big grin

FIGURE 5.1 *Continued*

>)................a sideways smile
:(..................a sideways frown
8-)................grin wearing glasses

SUBJECT HEADERS

Subject headers become very important when so many BIRDS roost together. Very clear warnings are needed when the material in a post could possibly offend. When a thread has veered off into new territory, the Subject Header should reflect the change—for example: What was "Re: My Dog" might become "Breakfast (was Re: My Dog)." This allows busy BIRDS to either delete what is to the reader an uninteresting thread or begin to read one that was being previously deleting. It just keeps things clear.

EDITING

Judicious editing is *very* important when replying to a post. This means cutting out everything from the previous post except the part you're answering, including all unnecessary headers. If you don't know how to do this, please ask! There are three main reasons why this is so important:

- Some BIRDS pay phone charges or access rates to download their mail and no one wants to pay for long posts full of unnecessary verbiage.
- None of us wants to have to page down through endless paragraphs to get to the gem you're adding to the thread. In fact, some may not bother and are known to get very cranky when there are loads of superfluous paragraphs to wade through.
- And besides, it's rude and you're quite liable to hear from someone telling you so, if you just hit "Reply to All" and type in a one-liner. Again, if you don't know how to use the "edit" function on your mailer, *ask*. There will be someone who uses the same one who will talk you through the process.

RESPECT

The single most important rule for happy roosting on BIRDS is *respect*. With over 300 of us, there's simply no way we're all going to agree on issues. And that's OK. But disagreements are *only* over issues and not personalities. There is no name-calling in this forest. "Flaming" is simply not allowed. We try very hard to remember that there's a real person on the other side of the monitor, and we treat him or her with as much respect as we want to be treated. Those who flame will lose their branch in the roost.

It's a good idea for those new to the forest to just perch for a while and read posts before flying off into the fray. Our conventions and ways of speaking to one another will soon become clear to you, but reading and "lurking" is about the only way one can get a feel for the place. Feel free to lurk for as long as you like. And when you have something to say, take a deep breath and jump right in. Let us know who you are. Tell us something about yourself. We're a friendly lot of birds, and we welcome new voices.

son's living room. Participants mixed fun with private self-disclosure in their conversation. The conversation lasted for hours with hundreds of real-time e-mails exchanged, some expressing enormous intimacy. In this and similar cases, some members expressed intense emotional and intellectual investment.

Interplay between Internet and Communication Principles

People who want to become more effective in their human communication on the Internet can benefit from understanding how the interplay operates between the Internet and human communication. In this section, we consider some generally accepted communication principles and adapt them to human communication on the Internet.

Internet Communication Is a Process

Communication is a continuous, ever-changing, ongoing, dynamic process. When you turn on your computer and connect to the Internet, electronic systems operate. Whether you receive new input, the electricity continues to flow and generate data in the system. At the point new information goes into the system from you or a source on the Internet, the system data enlarges, and it will never be exactly the same as it was before. Even if you shut down your system and start again, change results. Your computer system experiences use, wear, and effect. Human communication also continually changes as it builds upon Internet experience.

Adapting to the Internet Will Increase Communication Effectiveness

When you want to effectively run a program, you must abide by the rules and requirements of that program. It's rather like the rules of a game: You play by the rules or you can't play the game. Certainly there are instances when the programmer's intentions can be bypassed by the user, but basically the program provides structure for the Internet interaction. If, for example, the program gives an instruction that the user chooses to ignore, the program may not do what the user wants, or the user may be unable to predict outcomes. Sending messages to another person on the Internet works most efficiently if you adapt to the particular receiver. Is the receiver the program or another person? Sometimes one or both.

Communication Is Irreversible

Have you ever been in the situation where you got stuck online? You may feel caught or trapped within the program or web site and unable to manipulate it. Sometimes when an Internet user cannot achieve the desired results, the person may try to escape to a previous function, close a program, or disconnect. Once a person communicates with a web site or another person, that communication becomes irreversible. Your computer tracks your Internet activity as do many web sites. Your online discussion comment becomes part of the archive. One might say something regrettable, for example, but not even an apology can change the fact that the communication took place. At that point, the source experiences a loss of control and in the case of a discussion group, the only recourse may be to leave the group and move on to find new friends. That reaction is rather like "I don't want to play with you any more." If the two people are unable to mesh their experiences and understanding, it may result in a communication breakdown.

Internet Meanings Are in People

In language, people frequently credit words with specific meanings. Actually, the words themselves have no meanings, but the people who use them attribute significance, experience, and understanding to words. People often assume that others assign the same meanings to words, when actually they may not. Perhaps because machines are involved, one may expect communication via the Internet to be more exact, more precise, even mathematical. The Internet is closely aligned with science and mathematics, so communication should be precise, right? In the world of computer programs, yes. But often that precision causes other problems. Have you ever entered a URL address but made a mistake with one letter so that the web site did not come up? Of course. The computer is unforgiving and requires enormous precision. After all, it's rather dumb and cannot figure out what you meant, but simply interprets what you said. And when you communicate to another person on the Internet, then your message may be concise and lack the nonverbals that add so much meaning in face-to-face encounters. Figuring out meanings on the Internet becomes a kind of puzzle, sometimes a very difficult puzzle that you have to solve.

No One Can Experience Totally Effective Communication on the Internet

When a web designer has an idea, she or he begins the process of writing commands or using a program, but you may interact with the web site in ways the designer failed to anticipate. In branching, you may become stuck, distracted, misunderstand, or meet some other stumbling block. The chances of the web page working for you exactly the way the designer intended are remote. You may be playing or working on the web page in ways the designer never imagined. If you are communicating to someone via e-mail, that too is a difficult process fraught with potential misunderstandings. In face-to-face communication, we use nonverbals to understand the words. We look at the context, how the nonverbals reinforce meanings, contradict meanings, or interpret meanings. The limited nonverbals in Internet communication make effective communication a real challenge. So what do you do instead? Do you take the communication less seriously? Do you use more intrapersonal processing? Perhaps. You might work quite hard to figure out meanings. The self is a complicated and private influence on Internet communication, perhaps even more so than in face-to-face communication.

Effective communication skills can be learned. People are not born with face-to-face communication skills any more than they are born with Internet communication skills. Effective communication takes careful analysis, trial and error, reading, interpretation, and study. That learning process can be fun or work, easy or difficult, a joy or a frustration, but learning the interplay effectively can have valuable results.

Skill Section: Communication Design on the Web

Is simplicity more fun than complexity? Can complexity be made simple by design? Can design be playful? Professors of communication studies view writing differently from professors of English composition. Precise word choice, clarity through specific language, and simplicity of construct are key elements to effective writing according to the communication expert. So too, effective web design is not about glitz, but about communication.

Effective web design is not about glitz, but about communication. Users appreciate the elegance of simplicity. Pages need to load quickly and be accessible to the widest variety of end users. What looks exciting to the business executive hiring the web design job may not communicate effectively to a user. Further, research shows that people fall into several different cognitive styles, so the best sites appeal to various ways of knowing and processing information.

Text designed to be read online tends to be more readable in serif-free fonts, with print large enough to read on various screens. Text designed to be printed is most readable in serif font, with page size and layout appropriate for printing on a portrait-layout 8×11 piece of paper.

Many people want to download text or copy and paste text, so are thwarted when a designer uses graphic text or frames pages. Graphic text also is less clearly focused and more difficult to read than text, which can be an important point when designing easy navigation links. Many heavy Internet users quickly become impatient if they cannot load, search, and navigate almost instantly. One study, for example, found that the people browsing a site click an average of $1\frac{1}{2}$ times. That means users need to pull up the exact page they need with their first click.

Users seldom go online to see pretty pictures or to be taken through a guided tour with a Flash program going at the designer's favorite rate. With few exceptions, users want control and access. With millions of web sites available, users can quickly replace a slow or difficult-to-control web site with a site that is designed for effective communication. Consider these basic strategies for effective web design.

- Training sites tend to be quite linear in design. You want the user to gain specific information, so lead them through the information in a clearly logical way.
- Teaching sites are built around a strong narrative with opportunities for relevant digressions to stimulate the users' understanding, imagination, and application.
- Educational sites often are used by highly educated people. Avoid being too linear or restrictive. Users want "flexible, interactive, non-linear design."
- Reference sites need to pop up quickly and provide a way for users to easily find and print or download what they want.
- Provide the information the user wants in the fewest possible steps.
- Frames can be extremely frustrating to users. Frames-based pages behave differently from regular pages because they are really meta documents, not HTML documents. Forward and back functions, for example, fail to operate as users expect. More importantly, frames fail to operate on certain types of computers, certain sized screens, and computers designed for people with certain disabilities.

- Visuals slow down the loading process. Modest graphic content on a page takes about 10 seconds for the average viewer to load. Remove nonessential graphics, put them low on the page, or provide them as secondary access in order to improve the loading for your user's first view. Remember that many sites claiming to offer copyright-free graphics have actually illegally pirated the graphics. (Adapted from Lynch & Horton, 1997)

Reflection

What are some paradoxes of human interplay on the Internet? You probably have first-hand experience with Internet play, computer games, playful metaphors. But what paradoxes exist in this play of human communication on the Internet (see Table 5.3)?

The Internet intensifies playful interaction between people. We proposed these explanations of Internet play.

1. Internet play is for the self. It serves internal goals.
2. Internet play co-exists inside and outside the scope of ordinary life.
3. Internet play operates without fixed boundaries of time and space, although the play may operate within Internet rules.
4. Internet play is pliable. The Internet can completely absorb the player so as to integrate with "ordinary life" or take on a life of its own.

Do you agree? Are there more items we should add to the list? You may have experienced Internet play that is imaginative play, like a child would do. Internet play may intensify identity role-playing. If you talk online anonymously, for example, if you have pretended to have a different job than you really do, you are role-playing through the Internet. Does this kind of play seem light-hearted or dark-minded to you? What is your Internet interplay through fantasy, storytelling, computer games, metaphors, and online discussion?

TABLE 5.3 *Paradoxes of Internet Interplay*

Play.	Work.
Use cognitive collaboration.	Fail to think for oneself.
Connect with others.	Isolated and alone.
Play with new ideas for a more open-minded approach.	Polarize people.
Face fear.	Gain new and increased fear because of the overwhelming nature of information on the Internet.
Have individual fun.	Achieve global significance.
Enjoy free service to the Internet.	Invest enormous business and societal values, time, skills, and money.

Case for Discussion

In your course discussion group, give your analysis and perceptions on the following situation.

An exhibit by artist Nam June Paik circulated art museums around the United States a few years ago. The exhibit was entitled: "Electronic Super Highway," a name used to refer to the Internet. In Paik's exhibit, there were 500 television sets operating simultaneously, not to mention thousands of other pieces of technology from the electronic age. The portrayal of the Internet was fun, prompting some visitors to laugh out loud. Particularly interesting was the artist's representation of the World Wide Web, which was a room-sized display that surrounded the observer, looking much like a double helix. Perhaps he meant that the World Wide Web=Life.

In the Smith College Museum of Art Permanent Collection, http://www.smith.edu/museum/collections/sculpture, you can view Paik's "Internet Dweller." The human head in the art piece consists of two televisions as eyes, another television as the mouth, two clocks as ears, and the hair as lighted antennas pointed in every direction. What does this artwork imply about people who use the Internet?

One person who attended the exhibit said: "I liked the postal service information exhibit, which was empty and boring. That's how I feel about snail mail. Sometimes I don't check my snail mailbox for days, but I check my e-mail a couple times a day. My son's favorite exhibit was the real goldfish swimming inside a television set. The television is about watching life. The Internet is about participating in life. My favorite exhibit was the man made from electronic parts: He was life-size! His head was a television set, and his feet were computer keyboards. Makes sense because the computer is what our world stands on. The techno-man looked exactly like my husband, and he was even reclining in my husband's chair. No joke, the chair the same color and fabric as my husband's favorite chair!" To explore more about Paik's representations of technology, explore the Internet. ArtCyclopedia.com may be a good place to start: http://www.artcyclopedia.com/artists/paik_nam_june.html

1. How do you use the Internet to play?
2. Do you agree with this woman that our society stands on the computer?
3. What other interpretations of Nam Paik's technology art might be reasonable?
4. How does the artist's portrayal of the Internet provide new insights into your understanding of human communication on the Internet?

Internet Investigation

1. *Further Investigation of Play Theory.* Use a full-text database to look up the work of one of the scholars quoted in this chapter. Consider recent research, which has attempted to apply play theory to computer-mediated communication (Kuehn, 1993). Investigate theories that you can relate to play and communication (e.g., Csikszentmihalyl & Kubey, 1990). Find out what they said that is relevant to play on the Internet.

2. *Mass Communication View.* As you think about this chapter, how might you see the ideas through the eyes of an expert in mass communication? Investigate visual design people use on the Internet. Consider Avatar, which is sometimes associated with the occult. Avatar is the visual an individual uses for self-representation in a game or virtual environment. Investigate online and share what you find. You may want to begin at http://www.avatarpalace.net/
3. *Communication Design on the Web.* Design a web site that incorporates effective communication style. Your college probably offers space or you can use the course environment connected to this course. Include internal and external links. Incorporate the concept of play in your design and explain and defend your idea on the page you create. Is play simplicity? Humor? Elegance? As part of your investigation and design, find sites that you consider effective communicators. Describe any characteristics that make them playful. Investigate web sites that can improve your design, such as http://info.med.yale.edu/caim/manual/ Books to investigate include: Mok, C. (1996). *Designing business: multiple media, multiple disciplines.* San Jose: Adobe Press. Siegel, D. (1996). *Creating killer web sites.* Indianapolis: Hayden Books, www.killersites.com.
4. *Online Games.* Trace the history of games online, beginning with the games of the 1970s and continuing through a discussion of current games available online. Consider the multiple concepts of online games—e.g., gambling, age of game players, violence, characterizations, behavioral norms, role-playing, power, gender—in your investigation.

Concepts for Analysis

We have provided concepts relevant to this chapter so that you can check your understanding. These concepts are worth dissecting in more detail. As a research option, examine one concept. Use our citations as a starting point for your study. We recommend that you search for refereed articles in an online database. Find at least three solid references you can read. Summarize, scrutinize, and present your analysis to other people in your course.

Questions to stimulate your inquiry: Do you agree with the ideas as we presented them? What other points of view should be considered? What details need to be added to fully understand the concept? What can you contribute through your personal analysis of the concept? What more do scholars need to know about the topic?

Computer pscience is a concept about the play of computer communication. The p is silent just like the p in the word psychology. The p recognizes the way computer use creates internal human processing simultaneously with the electronic computer processing.

The idea of *fantasy themes* is an idea suggested by Bormann (1980), who thought that by sharing stories, individuals create a group fantasy that describes their group culture. These fantasy themes tell much about the people involved in family, organization, or group. In fact, the fantasies of the storytelling are probably more meaningful and important than the reality of what actually happened. The Internet is the ultimate group fantasy: The Virtual World.

Interplay refers to interaction, which can be entertaining, interesting, or lighthearted.

IT is an abbreviation for Information Technology, which is an inclusive term meaning all types of electronic hardware, software, and processes designed for message transmission, data storage, and information calculation.

A *metaphor* is a symbol used to represent something it is not. The Internet can be used as a metaphorical model of the human communication process. A *metaphoric word* is one that draws an analogy or a comparison. The word Web, for example, brings the image of a spider on a web.

MOOs, and MUDs are multi-user dimensions. These Internet user groups exist for educational or entertainment purposes and often develop elaborate rules of interaction. Often the users employ significant fantasy in the interaction, even pretending to be together in a real place or pretending to be certain characters.

Play is having pleasure, fun, and enjoyment through Internet communication. The play includes four types of play: (a) Imaginative play like a child. (b) Identity role-playing. (c) Interplay with the self or other. (d) Play that actors act in.

Symbolism is the representation of meaning through words, actions, or other means.

Synchronous interaction happens in the same time. For example, two people e-mail each other when they are online at the same time and e-mailing each other back and forth.

A *Web page* is a single location—URL—on the Internet.

Notes

6. http://www.sci.tamucc.edu/~pmichaud/grape/
7. http://www.sci.tamucc.edu/~pmichaud/toast/
8. http://www.meowmail.com/
9. It was only after deliberation that we chose to keep this example in the book, recognizing that some people will find it offensive. We apologize to them, but we wanted to represent the sexualized side of the Internet.

6

Polarization of People

> *"Arab schoolchildren in Jersey City told their classmates ahead of time that the World Trade Center would be attacked."—Pure hatemongering. This statement is an example of the kind of Internet rumors that circulated after the September 11, 2001 attack.*
>
> —(Filkins, 2001, p. B8)

> *"Congress is about to impose a 5-cent surcharge on each e-mail message you send!"—Bald-faced lie.*
>
> —(Bowen, 2000, p. 124)

> *"Cooking in aluminum pots causes Alzheimer's disease."—Wrong.*
>
> —(Brody, 2000, F. 8)

Does Internet communication enable people to find common ground or intensify differences? Both.

Think about the evolution of technology in this country. A pattern that has been consistent in the history of communication technology is that with each new medium of communication, the same social battles intensify between people with differing interests in the effects of the media (Czitrom, 1982). When people are in crisis, they tend to revert to their old, familiar, and narrow ways of thinking instead of the open creativity that enables them to solve problems effectively. The same reversion happens in societies during periods of crisis as they take away freedoms because of the fear of the potential for change that comes from freedoms. "Change creates anxiety—and anxiety brings with it behaviors that stem from fear and insecurity" (Barclay & Wakabayashi, 2000, para. 13).

From a distance, this conflict seems to be between those who want the new medium to develop because it appears to have so much to offer and those who want to suppress it because it threatens our society. So far, evidence rebukes both extremes (Phipps &

Merisotis, 1999). The actual state of affairs often reflects both positions simultaneously: The idea that Internet-mediated communication is a new frontier with opportunity awaiting and also a dangerous mine field about to do harm to innocents who tread upon its terrain. The Internet is good—use it! The Internet is bad—stay away from it! This double bind extends beyond simply whether to use the Internet, but how to use the Internet, and with whom to associate via the Internet. Indeed, some people find the Internet to be a safe-haven for arrogance, hate, and an intensification of differences.

Focus

TO: Students

FROM: paradoxdoc@hotmail.com

RE: Overview: Polarization

In contrast to the playful intensity discussed elsewhere, here Lenny and Joan discuss the frightening reality of how people become polarized through the Internet. They debunk e-mail myths, which users tend to believe simply because they are in print.

The Internet may actually polarize people within a home. They discuss how people in discussion groups often clash with the ideas or people you encounter. When uncomfortable, those people can keep trying different groups until they find just the right bunch of people who think like them. So, even people who are outside of mainstream thought can find people scattered all over the country who think in ways very similar to the way they think. Further, information sources can be geared to individuals so that they receive information consistent with their predispositions.

Lenny and Joan ask you: Does the Internet increase democratization or create a digital divide? Communicating through the computer can help people make or meet their enemies. By examining such concepts as cognitive dissonance theory and the use of video games, Lenny and Joan ask you think about many questionable aspects of Internet communication. The gain or loss of privacy and the circulation of hoaxes, rumors, and myths are two such Internet communication problems.

> Lenny and Joan explore elements that polarize people through the Internet. They also mull over other parts of the Internet's dark side—Internet addiction, hostile metaphors, flaming, personal fear, hate speech, and terrorism—are also discussed.
>
> Perhaps the "perception that the world is a dangerous place" fuels the hostilities often expressed on the Internet. Some people who choose to interact with the world through a computer instead of more traditional face-to-face methods may be using the computer as a buffer against what they consider a cruel world. Individuals may hide in the protection of their homes instead of venturing out to interact with people face-to-face. The violence experienced online can seem like a surreal or unique reality. When there are real attacks on individuals—which are intensified by the Internet, such as stalking e-mails and harassing web sites—individuals or groups of people may understandably feel extremely threatened. In contrast, some people may use the Internet because it allows them to face their fears.

In this chapter we will explore the rhetoric that uses the Internet to polarize people. By *polarize,* we mean that the Internet encourages groups of people to emphasize their differences with others so they are repelled by other people. At the end of this chapter, you should be able to answer the following questions.

1. How does the Internet increase polarization?
2. Does the Internet create a digital divide or the democratization of culture?
3. How does cognitive dissonance relate to Internet interaction?
4. What hostilities are expressed through metaphoric language?
5. What is the nature of hate speech on the Internet?
6. What is the role of video games?
7. Does the lack of privacy increase your fears of Internet use?
8. What is an Internet hoax?
9. Why are rumors and myths so credible on the Internet?
10. Are you afraid of becoming addicted to the Internet?
11. What are some kinds of fears generated by the Internet?
12. How does the Internet fuel fears?
13. How is the Internet vulnerable to attack?

14. How do people use the Internet to face their fears?
15. What polarizing paradoxes have emerged through communication on the Internet?

Intensification of Polarization

The Internet may actually polarize people within a home. One woman told us that she had developed marital difficulties because her husband wouldn't get off the computer. Of course, you might wonder whether the computer is a problem or a symptom of a problem. A mother of four told us that their home computer was a source of conflict. The children "fight to get on it. You'd like to use the computer for an extended length of time, but someone else always wants a turn." People who enjoy using the Internet are frustrated by the slow speed and interruptions of their connection during the day. An advertisement for a direct Internet connection company portrays the same kind of problem, where each family member schedules a time to be online in the middle of the night because it is so hard to connect during the day.

In this chapter, we are more concerned about the way people connect with others to increase polarization of people. If you look around your neighborhood, for example, perhaps there are 100 people available with whom you can interact with the potential of becoming friends. Frankly, you probably don't like the looks of some of those people (let's estimate 25). Maybe another 25 are outside the age range of people you want to talk to. Maybe another 25 think too differently from you, aren't educated in the same way, or are too different for your tastes. That leaves you 25 people with whom you will talk and make friends. Of course, the selection of numbers in this example is arbitrary simply for the sake of illustration.

Now consider your Internet neighborhood. There are millions of people available, with whom you can interact. Frankly, you can't see what any of them look like, so you can't screen them that way. Nor can you tell immediately how old they are. That leaves you millions of people with whom you can still interact. Time and space have different meaning online. If you have insomnia, for example, you can become e-pals with people on the other side of the world. So, how do you decide with whom you'll talk? You have thousands of chat rooms and discussion groups available, so you start selecting by criteria such as politics, where you went to college, interests, religion, gender topic, hobby, age, sexual orientation, geographical region, language, nationality, or whatever else you choose.

As you try out a certain *discussion group,* you may find that you clash with some of the ideas or people you encounter. You could keep trying different groups until you find just the right bunch of people who think like you. So, you find people scattered all over the country who think in ways very similar to the way you think. Imagine that someone mentions the Yahoo page where you can put in information about yourself so you'll give just the right slant on information. You could be very specific giving your age, political affiliation, and employment. Perhaps some of your computer information has been given through cookies so that the Yahoo portal can track your online activity, such as where you shop online, what web pages you look at, what discussion groups you join. The Yahoo computer analyzes all that information and gears your news, information, and entertainment information directly to who you are.

Are you beginning to understand how the Internet can bring together people who think alike? How does the Internet reinforce entrenched attitudes, beliefs, behaviors, and values? How can the Internet actually separate people so they become more polarized? First, the vast numbers of people online permit people to connect with people who are very similar to themselves. Second, information sources can be geared to individuals so that they receive information consistent with their predispositions.

> Due to the global scale of this technology and the fast pace of innovation, computer-mediated communication (CMC) has a clear potential to breach boundaries of nationality, race, language, and ideology (e.g., Hiltz & Turoff, 1978). Indeed, this connectivity has led many commentators to speculate about the breakdown of (traditional) social boundaries, the implication being that we are now just all individuals in the "global village." Enhanced communication allows us to traverse, and thus potentially transcend, social boundaries by facilitating the proliferation of both standardization across, and social differences within, communities. On the other hand, the new communication technologies also provide the prospects for developing new ("virtual") communities and social identities, thereby erecting new boundaries as well as breaking down old ones. The implicit assumption of both these analyses, however, is that electronic communication will help to crosscut traditional boundaries and undermine the bases of social division in its many senses. The utopian image sometimes conveyed is that the new technology affords a new and more liberated way of being. (Postmes, Spears, & Lea, 1998, para 3)

We see the Internet as a communication medium of paradoxes.

> A group—or, more generally, any social category—exerts on its members an influence that restricts and restrains behavior, but more positively can also be seen as a source of (social) identity and self-expression. Different boundaries are thereby imposed on group members, but group members also impose these boundaries on themselves. Group norms and social stereotypes define the limits of social behavior that are often used to differentiate groups to which we belong from those to which we do not. Social boundaries define where the ingroup ends and the outgroup begins, and what is appropriate conduct within the intragroup and intergroup context. (Postmes, Spears, & Lea, 1998, para 4)

In the rest of this chapter, we will consider cognitive dissonance, hostile metaphors, hate speech, video games, and other parts of the dark side of communication on the Internet. Remember, there are many paradoxes about the Internet. Yes, the Internet can connect people, bringing them together in some ways. And yes, the Internet can also divide and polarize people.

Digital Divide or Technical Together?

Consider the role of the Internet in dividing a family we interviewed. Their responses indicated that the computer was a divisive force in this family. Both father and son said they frequently felt angry while using the computer. The father said he talked to other family members less since he purchased the computer, and both the father and son said the overall effect of the computer on their family was that it actually had driven them apart. The family indicated severe problems in the interpersonal interaction among family members. One must question, of course, whether the computer caused the communication breakdowns, or simply manifested their poor interpersonal communication. Whether or not a direct relationship exists, the family demonstrated a negative association between their computers and their family communication.

In contrast, another family we interviewed considered the computer to be a force that brought them together. All family members agreed that using the computer was fun. They also agreed that the computer prompted positive feelings of satisfaction and intelligence and the play function was essential. The father thought the computer actually brought the family closer together, and the whole family agreed that they shared their computer as a hobby and way of playing games together. The family associated the computer with positive interpersonal communication.

But what about on a societal scale, does the Internet bring people together or pull them apart? Papert (1981) theorized that with the computer would come a greater division of classes. The socioeconomic dichotomy would increase due to lopsided computer skills: "As knowing how to use a computer becomes increasingly necessary to effective social and economic participation, the position of the underprivileged could worsen, and the computer could exacerbate existing class distinctions" (pp. 86–88). Some people believe that the Internet creates a *digital divide* because only the advantaged can have and use the Internet effectively (Jung, Qiu, & Kim, 2001). The digital divide concept says that there is inequality regarding Internet access, use, search strategy knowledge, the variety of uses of the Internet, the speed and quality of technical connections, one's use of social support, and ability to evaluate the reliability of the Internet (Anderson, Bikson, Law, & Mitchell, 1995). In fact, strategies such as complicated web sites that require DSL-type connections, walled gardens such as the one operated by AOL, large software programs that require large capacity computers, subscription services, and similar procedures make the Internet accessible only to people who have resources. Thus, the Internet creates a greater difference between the haves and have-nots. One study, for example, found that status differences existed in electronic groups and high-status individuals dominated discussion (e.g., Weisband, Schneider, & Connolly, 1995). The idea is that the Internet provides the acquisition and advancement of knowledge, but in a way only available to some people. For example, many people are still not connected to the Internet in this country. Residents of non-urban areas, those living in southern states, and people with less education are least likely to have Internet access. Households with school-aged children are most likely to have Internet access (Wellner, 2001). Thus, people in our country have access to information resources not available in every part of the world. Further, as you might expect, 97 percent of the host computers of Internet sites are located in developed countries so that people in developing countries may have quite limited access to the Internet (DiMaggio, 2001).

Other scholars have argued that the Internet democratizes United States society and cements people within our society (Hoffman & Novak, 1998; Jackson, 1999b; Schofield, 1997). The *democratization* ability of the Internet has been touted by scholars (e.g., Dahlgren, 2000; Netchaeva, 2002) because computer-mediated communication reduces communication and status differences (Wajcman, 1991). "A NEW nonliterary culture exists today, of whose existence, not to mention significance, most literary intellectuals are entirely unaware," wrote Susan Sontag in her groundbreaking 1966 essay, "One Culture and the New Sensibility." The new sensibility, she argued, collapsed the distinction between highbrow and lowbrow, embraced popular culture, celebrated modernist music and painting the masses had little taste for, and advanced a new understanding of the senses (Skinner, 2000, para. 1). We could say that the Internet has moved us farther in that direction, and thus closer together.

Today, a person can communicate effectively on the web, including designing web sites with little money or expertise in technology, and can communicate to a mass audience. Internet discussion groups have become a type of mass movement, providing the ability to communicate daily with people from all over the world. The Internet gives easy access to information, so that everyone has equal opportunities.

Freedom to access information, however, is not necessarily emancipation. Internet advertisers tend to use a divide and flatter approach that goes after the young, non-product-loyal consumer who is most likely to be influenced. Often discussion group freedom is used to flame, belittle, and squelch certain voices, often voices of older women and minorities (Scodari, 1998). As the Internet is more widely adopted, however, these differences are subsiding (Compaine, 2000). *Portals* have helped solve some of the problems of digital divide. Many web sites provide comprehensive services, including search engines, category guides, evaluations of the credibility of other information sites, and services, which can be quite easy for the novice to use. The objective of these portals is to make the Internet easier to use so that people can access the information they need.

Cognitive Dissonance

Computer use affects all people within our society because we have easy access to information and people around the globe. We can use computers to communicate to people who think like us and ones who hate us even though they've never met us. Communicating through the computer can help us make or meet our enemies. In other words, the Internet can intensify the polarization of people. Remember, by polarization, we mean that people who are similar cluster together, reinforcing each other's similarities while standing against people who are different. In some cases, this polarization may be benign, in other cases, the polarization may be part of insidious hate and violence.

As people come to replace their local newspaper with information portals on the Internet, they can select filters that adapt to their interests and biases. The Internet can enable people to access the world news seen through filters that can intensify their attitudes. We can find an explanation for this phenomenon in *cognitive dissonance theory* (Brehm & Cohen, 1962; Festinger, 1957; Petty, Ostrom, & Brock, 1981). The theory of cognitive dissonance suggests that our beliefs or cognitions can be related either by their consistency (consonance) their contradiction (dissonance) or their irrelevance (lack of connection). In general terms, we tend to hear and seek information that is consistent with our beliefs. We tend to avoid or refuse to believe information that is inconsistent with our beliefs (dissonant). The number and importance of consonant and dissonant elements within a cognitive set determines the amount of dissonance we feel when challenged. When we experience significant dissonance, we have to reorganize our beliefs or somehow dismiss the value of the source of dissonance.

Cognitive dissonance appears particularly important in understanding the Internet. First, through the Internet, people come into contact with people and ideas they would never face under normal, onground, face-to-face circumstances. Differences in background, ethnicity, education, economic level, and other factors create a heterogeneous mix of people who interact on the Internet. This mix of people means beliefs are frequently challenged by unfamiliar information and arguments.

Second, people often hold inconsistent beliefs, but experience no dissonance because they are unaware of their inconsistencies. In an online discussion, those inconsistencies may be brought to an individual's attention, causing serious dissonance. The dissonance becomes unavoidable and can feel seriously threatening, resulting in intense hostilities.

Third, the public nature of the Internet intensifies reactions. An individual may say something to a family member in private, and the comment can be dismissed. The same comment made in an Internet discussion might be read by dozens of people and warrant all kinds of interpretations and responses. The public nature of inconsistencies arouses dissonance (Brehm & Cohen, 1962). Suddenly the discussion participant may feel confused, defensive, as the person's belief system is threatened by the online discussion. This kind of behavior is what seems to prompt many participants to storm out of a discussion group.

Fourth, while some group participants may be seeking new challenges, most are probably seeking confirmation of the beliefs they already hold. Instead of finding people who will expand their thinking, they probably move from group to group to find people who have like minds, who help solidify their already held beliefs. Instead of moving people toward a common ground, the Internet moves these people toward polarization.

Video Games

> The cover features a young coed in a revealing negligee. She's screaming. Behind her, Ninja-like intruders in black costumes creep forward. They've got hooks that will clamp around the young coed and a drill that will bore into her neck to stop the screaming for good.
>
> It sounds like a bad B-movie. It's actually the Sega video game Night Trap. (Schroeder, 1996. para. 1, 2)

How do Internet video games intensify the polarization of people? Parents, politicians, religious leaders, and teachers have expressed concern over the negative effects on attitudes, behavior, and self-concept associated with people who play violent computer games (Funk & Buchman, 1996). Each year some new games are released for adults only. Do you think violent games have negative effects? Have you ever played a violent game online?

Jonathan Cowan of Americans for Gun Safety expressed concern on an ABC news interview (2003). There is a new level of violence and realism in current video games. In the death of one 13-year-old, the children were acting out a scene in a specific video game, which resulted in a fatal stabbing. In the game Max Payne, the player uses a handgun to shoot people and police cars. In Redneck Rampage, the player drives drunk and shoots at innocent people along the road. Of course, just because the behaviors are associated, that doesn't mean they are causal. In other words, people with aggressive behavior may be drawn to playing violent computer games, whether or not the games increase their aggression. But, the interactive nature of the person playing a video game, whereby the person becomes the violent actor, may have a profoundly negative effect on certain people and a generally negative effect on the average person.

> The prevalence of game play indicates that it is unlikely that playing video or computer games causes severe psychopathology in the average player. Related media research suggests, however, that frequent exposure to violent electronic games may have a subtle negative influence over the long term and may decrease empathy, disinhibit aggressive responses, and strengthen the general perception that the world is a dangerous place. (Funk & Buchman, 1996, "Game Playing," para. 2)

Privacy

This book has already discussed how information is collected about you through your Internet activity. Some people are threatened by the potential loss of privacy through their Internet use. You probably know cases where child predators, corporate crooks, dishonest hackers, and individual identity thieves have taken advantage of the Internet. Many people feel a sense of anonymity online, as if they are free to explore whatever Internet sites they want, to express whatever they think in e-mails, believing that others will not know who they are, so they cannot be held responsible for what they do and say. Other people are careful to use anonymous remailers, firewalls, virus protection software, encryption software, and other strategies to protect themselves online (Gelman & McCandlish, 1998).

Savvy users understand that servers that connect on the Internet can store e-mails and software can examine their content. A server can track your web site searching, take cookies to build a profile of an individual, which can be used to gear advertising spam. An employer can use monitoring software to keep track of your Internet behavior. Your neighbor or a person across the world can try to hack into your home computer if you share certain Internet services. A hacker can steal or destroy files of the work you keep on a server. Thus, another paradox is that the Internet increases freedom through access while decreasing individual privacy.

> Without question, the growth of government and commercial transactions and the increase in technological developments over the last 50 years have heightened threats to privacy. Today the Internet accelerates the trend toward increased information collection and facilitates unprecedented flows of personal information. Cellular telephones and other wireless communication technologies generate information about an individual's location and movements in a manner not possible until now. Electronic communication systems generate vast quantities of transactional data that can be readily collected and analyzed. And law enforcement agencies, particularly at the federal level, place increasing emphasis on electronic surveillance.

Confronted by these challenges, there are still grounds for optimism. While dangers to privacy capture our attention, they sometimes lead us to understate the unprecedented gains in privacy protection that have also been achieved over the last half of the twentieth century. In many cases the legal system has laid a foundation for privacy protection through court decisions, state and federal legislation, and self-regulation. (Berman & Bruening, 2001, para. 2–3)

Hoaxes, Rumors, and Myths

An Internet *hoax* is an elaborate practical joke which conveys false information through the Internet. An *urban legend* is a type of hoax which spreads a myth. Perhaps Internet hoaxes are designed to be an extension of the human desire for humor and an appreciation for practical jokes. Urban legends "serve as a psychological shield against the unknown and cause no real harm themselves" (Harmon, 10/12/2001, B. 12). "These spammers email their bogus tales to as many recipients as they can think of in hopes that someone will be ignorant enough to believe them. These recipients unwittingly continue the cycle of misinformation by forwarding the bogus message to their friends and loved ones" ("Fight hoaxters," 2000, p. 14). One of the problems with this spam is clogging up the Internet bandwidth, cluttering mailboxes, slowing connection speeds. Further, the spam can create misinformation, fear, and panic. There are many kinds of Internet rumors and misinformation (Bordia, 1996). And this misinformation includes myths including bogus virus warnings, business and product boycotts, threats, ailments, chain letters, and hate mongering ("Fight hoaxters," 2000, p. 14).

Thousands of urban legends circulate on the Internet (Brunyand, 2001). The speed, availability, and pervasiveness of the Internet provide an amazing vehicle for gossip, yarns, inflammatory statements, rumors, myths, and bald-faced lies. Internet urban legends create hope and hate. Each individual can be a "national broadcaster" with e-mail and web pages. First, U.S. citizens have long valued the printed word as somehow more credible than the spoken word. Second, we tend to believe information that comes from mass media, then we can believe that information; web pages from valid news sources look much like web pages from individuals. Third, we expect U.S. businesses to provide valid information. When an e-mail address is an esteemed U.S. business or university, we think that message must be valid. In fact, the e-mail is sent without anyone in authority knowing about the e-mail. Fourth, repetition can make the fantastic seem more plausible.

> Internet rumormongers recirculate the same small set of stories. But the more that tales are shuttled between in-boxes, the more they acquire a false aura of truth by virtue of sheer familiarity. These messages gain verisimilitude from the silent authority of written text, which can seem persuasive when a similar message, expressed in pictures, would be instantly dismissed as false. (Stross, 2001, p. 44)

This undeserved validity makes misinformation via e-mail and the Internet a serious problem. Some people react to cybermyths with fear and panic, particularly ones related to health and terrorism (Brody, 2000). Thus, with the credibility of the written word, the

Internet hoaxes take on a plausibility and credibility for the most ridiculous theories and rumors. For one thing, the rumors are often repeated over and over, making them appear to come from a variety of credible sources. (Filkins, 2001, p. B8).

After the September 11, 2001 attacks, Nostradamus, sex, and mp3 were among the most frequently used terms in Internet search engines. According to circulating e-mails, the French astrologer said that "the third big war will begin when the big city is burning" after "two brothers" are "torn apart by Chaos." In truth, steel didn't exist at the time the astrologer supposedly warned about the "steel birds," and neither did the astrologer—the quotations were dated after Nostradamus' death (Harmon, 9/23/2001). Another false Internet rumor was one about an Afghan woman whose boyfriend warned her not to fly on September 11 and not to go to malls on October 31, and other false reports about people of Arab descent. These false messages have been called a "cultural epidemic" and "thought contagion." Messages that feed on fear spread more quickly on the Internet (Kakutani, 2001). The Internet is a medium of contradictions. In the same way the Internet allows some people to perceive themselves as more similar than they might understand face-to-face, so too can the Internet allow people to intensify their differences. We are not suggesting that the Internet increases paranoia or unrealistic fear. In fact, the Internet may enable people to understand the risks of human communication.

Internet Addiction

The idea of *Internet addiction* is that some people cannot control their behaviors, so they use the Internet to the extent that the Internet interferes with normal activities. One of us participated in an online discussion group of psychology and communication studies scholars. Periodically, someone would come into the e-group and raise the concept of online addiction. Then at least one member would tell how the whole concept started as a joke. The psychologist involved coined the term "Internet addiction" to show how people would grab onto nearly any concept as something that needs a cure. In fact, the psychologist explained, the person who uses the Internet extensively does so to work or communicate with others.

If the person spends entertainment time in a discussion group, the individual is doing much the same kind of communication she or he might have at a dinner party. We have no "dinner party addicts," although there are people who enjoy having guests to dinner for conversation. We have no "get-my-job-done-with-clear-focus" addicts, yet people who do just that in an online environment are sometimes accused of being addicts. Certainly there are people who manifest unhealthy and extreme behavior on the Internet. For example, there is evidence that some people obsessed with sex will turn to the Internet. And our society has psychopaths and criminals who are facilitated through the Internet. But these people are disturbed, and the Internet is a manifestation of their disturbed mental state and asocial behaviors. With that said, we should say that some scholars take the concept of Internet addiction quite seriously (e.g., Young, 1998).

We are not psychologists, so we chose not to enter this debate. We are researchers exploring communication on the Internet. Although there are communication behaviors which have negative effects, we see the process of communication as a necessary part of humanity, and healthy communication behaviors can be part of the Internet as well as any other type of human communication. We see no difference in someone checking e-mail for a message from a colleague and calling the colleague on the phone, or walking over to his office to talk. The difference may be, however, that the wasted time and effort of checking and finding no e-mail takes far less time than the phone call or walk to an empty office.

For the people who prefer to talk to people on the Internet and have no face-to-face relationships—obviously, we've never met anyone like that—we are glad they have managed to find a way to connect with others, even if it is "just" by the Internet. And there is some evidence that suggests people can learn something about creating and maintaining relationships by their Internet interactions.

For the people who moan that their spouses are always on the Internet instead of willing to talk to them or go somewhere, we don't see any difference in that behavior and spouses who sit glued to the television. Actually, we do see a difference. The Internet person probably interacts more effectively mentally than the television viewer. The real problems here are the poor interpersonal relationships reflected by avoidance behavior; the problem is not with the Internet.

We do see another paradox: The Internet may attract compulsive behavior, but the Internet can open opportunities for people to become more normal (Barnes, 2002, p. 196). But our argument sounds a bit like "Guns don't kill people, people kill people." Truth is, the Internet doesn't destroy lives or marriages or relationships; people do that. Does the Internet intensify negative effects? One researcher, for example, blames "the Internet for people's "interpersonal, social, and psychological problems" (Young, 1998). We disagree with Young. Negatives may be intensified on the Internet, but positives can be intensified too. The Internet does intensify the individual's interplay with the cognitive self, collaboration with others, learning, and diverse ideas. Internet interplay may intensify the individual's understanding of the lack of interplay in her or his physical environment. And the Internet may intensify the individual's intrigue with her or his cognitive interplay with others.

Hostile Metaphors

Those who study communication have long believed that our language affects the way we understand our environment (Carroll, 1956). If we examine the language of the Internet, we can find many examples of metaphorical language which ultimately affect the way people attach meaning. A *metaphor* is a word used to represent something it is not, a symbol that draws an analogy or a comparison to give a precise meaning. *Netspeak* is a new language dimension, that is, computer-mediated language is unique through its elaboration and enrichment of conventional language (Crystal, 2001). One possible area of influence of Netspeak is in the use of hostile metaphors in online communication. We do not suggest that these are the only types of metaphors used online, but many of the online metaphors appear sexist or violent.

People who create and edit web sites are often called webmasters.

According to Bill Cullifer, executive director of the World Organization of Webmasters (WOW) in Fulsom, Calif., "the word originally came from postmaster. The early

Webmasters were those academics who routed electronic stuff." ... But master, from the Latin magister, "chief," has always been associated with males, especially controlling, dominant, insensitive fellows. Its female counterpart is mistress, sometimes pronounced "gerfren." (Safire, 1999, para. 15–16)

Now consider how one web editor feels about the term webmaster.

Everytime I see the word "WebMASTER" I wonder at the audacity (or ignorance) of the engineer who dreamt up that word. I just can't understand why a person who edits a web page is not an editor but a "master."
 Was it intended to be part of the male mystique to make sure everyone knew that Internet was a male domain—or was it an accident of language by English-language challenged engineers whose vocabularies did not include words beyond bitch, tit and. ... Everytime I see that expression "webMASTER" I can see some computer nerd snapping his keys like an imaginary whip—let's get those "nigger" words beat into shape, men! We're the masters!"
 I am a webitor, a web editor, or an editor of a web site. And I refuse to acknowledge the term webmaster—the MASTER of all HE surveys. (Irene Stuber http://www.undelete.org/)

Among the many metaphors used in language about computers and the Internet are terms often associated with violence (e.g., blow away, brute force, locked and loaded, search and destroy mode). Research has suggested that perceived language variations trigger evaluative judgments about the source (Strand, 1999), particularly when involving stereotypes and prejudicial words (von Hippel, Sekaquaptewa, & Vargas, 1995). Metaphors can exert power because they suggest fact (Burke, 1969, 1984). While shaping our perceptions, we use metaphors without giving them much thought (Lakoff & Johnson, 1980, p. 3). Although Internet metaphors contain positive and negative meanings, there seem to be many terms with negative connotations. Parks, Roberton, and Safire (2000) called sexist language: "words, phrases, and expressions that unnecessarily differentiate between females and males or exclude, trivialize or diminish either gender" (para. 1). Some Internet metaphors appear sexist, which is significant because of their potential effects. Research has suggested negative effects from such language because the language affects perceptions (Bing, 1992), suggests male superiority (Gastil, 1990), and can affect self-concept (MacKay, 1980). Consider the sample of metaphoric language from the Internet shown in Table 6.1.
 Given the reviews of research that have suggested that men have dominated women in technology (e.g., De Palma, 2001; Whitley, 1997), one cannot help but wonder if Internet language has contributed to an oppression. In early computing, women dominated computer operations. Then there seemed to be a period when males had higher skills and better attitudes toward computer use, according to Whitley's meta-analysis. Confirming the perception of male domination, Janssen, Reinen, and Plomp (1997) studied educational computer use around the globe and found that males had more skills and enjoyment than females. Other studies suggest this trend is reversing, however, and in fact, computer use might be a highly feminine activity (e.g., Camp, 1997; Newton & Beck, 1993).
 Are violent Internet metaphors designed to oppress women and minorities? Content analysis of the English language has suggested that there are more male-valued than

TABLE 6.1 *Internet Metaphors*

Word	Metaphor or Potential Secondary Connotations	Definition
black screen of death	violence	Failure mode in Windows.
blast	violence	Ruin, also nuke.
blow away	violence	Accidentally remove files and directories.
blow up	violence	Become unsafe.
bondage-and-discipline language	violence	A language supportive of the author's theory.
brute force	violence	Primitive programming style.
dark-side hacker	violence	Hackers that form a technological elite—Jedi Knights who are seduced by the dark side. Criminal or malicious hacker; a cracker.
deadly embrace	violence	Deadlock, neither process can proceed because each is waiting for the other. There are many words using dead or death (e.g., DEADBEEF, dead link, death code, death star).
decay	physical	Automatic conversation from one to another.
demon or daemon	devil	Processes in a program that lie dormant until initiated.
dickless workstation	sex, violence	Highly server dependent.
examining the entrails	bodily	A type of process looking for a program or system bug.
gang bang	sex, violence	Fast software development by a group of people.
list-bomb	violence	Subscribing a person without permission to mailing lists so they receive excessive mail.
lobotomy	physical, body metaphor	Management training of a hacker.
locked and loaded	military	Removable disk locked into the drive, the heads loaded.
logic bomb	violence	Related to code designed to cause destructive or security-compromising activity.

TABLE 6.1 *Continued*

Word	Metaphor or Potential Secondary Connotations	Definition
mail bomb	violence	To send, or urge others to send, massive amounts of e-mail to a single system or person, esp. with intent to crash or spam the recipient's system.
war dialer	military	Telecommunication dialing cracking tool.
war lording	military	Excessive signature.

female-valued terms in the language (Sankis & Widiger, 1999) and that tendency probably carries over into these kinds of Internet metaphors. Will computer users move toward more gender-neutral terminology? According to Safire (1999), although advocates of gender neutrality have logic on their side, many people find problems in using nonsexist language and continue to resist using nonsexist language. Internet language may qualify as "subtle sexism" (Boggs & Jansma, 1999) because it is accepted, ignored, and considered inconsequential (Benokraitis, 1999, p. 11).

The style of interaction on the Internet is North American/European dominant, but people have created their own culture-specific style. Talk in the Internet culture often uses a Western attack-and-defend approach. People can "sound" blunt and aggressive on the Internet, even when they do not feel that way. The brevity and haste of Internet messages can lead to misimpressions. In one study, for example, respondents were more likely to answer honestly to sensitive, softened questions online rather than blunt questions (Peiris, Gregor & Alm, 2000).

Flaming

In contrast, people often intend to sound blunt and hostile on the Internet. *Flaming* is when people exchange hostile or insulting remarks. Flaming can create problems in e-mail conversations (Thompsen, 1994), and we have observed how the behavior is destructive to the group climate.

What Stimulates Flaming?

The usual explanation for flaming is that the computer barrier and distance make some people feel more justified in verbal abuse of others. As with telephone communication, there are fewer consequences to being rude online in contrast to face-to-face interaction.

Perhaps flaming is simply an outlet for expression. If a person does not know how to deal with aggression toward discussion contrary to her or his beliefs and values, the participant may lash out at others in the group. We have noticed some people seem more willing to discuss controversial ideas online than face-to-face, and without those ideas couched in the positive nonverbals they would have used face-to-face, some serious conflicts can arise. Some people may not realize their words are perceived as attacking the person instead of the idea. Further, some participants seem so emotionally involved in conversation on the Internet that they react more strongly than they would in a typical face-to-face situation. Sometimes trolls—people who make outrageous comments—act as pranksters to disrupt the group, gain attention, or make new individuals feel clueless (Barnes, 2002). Other individuals may use flaming as a control strategy when seeking group leadership (Mabry & Thomas, 1991).

Although chat rooms, discussion groups, and messageboards usually are intended to be a place to give people of like-minds a place to talk, exchange ideas, and give support, people there may be the target of hate flames. Although some service providers monitor chat room and discussion group talk and have rules against abusive behavior, the monitoring doesn't always work. "Allah must die" and "All muslims and jews are terrorist" were among the hundreds of abusive comments observed in AOL's Islam-related chat rooms immediately following the 9-11-01 attack on the United States (Harmon, 9/23/2001).

Perhaps some computer users are overly controlled and nonassertive in face-to-face communication, so they unleash their aggression in flaming on the Internet. Still another explanation for flaming, however, is that flaming is a way of making interactants feel more intensity in a medium that lacks the direct interplay of human senses. In this case, argument makes the communication seem more real and lively. Regardless of why a person uses flaming, you can chose how to respond. You might ignore the posts. You might find the argument stimulating and challenging. You might unsubscribe from a group. You might install a bozo filter. Even online, we decide how to respond to other people, and we decide how other people will respond to us.

Personal Fear

Perhaps the "perception that the world is a dangerous place" fuels the hostilities often expressed on the Internet. Some people who choose to interact with the world through a computer instead of more traditional face-to-face methods may be using the computer as a buffer against what they consider a cruel world. The individual may hide in the protection of her or his home instead of venturing out to interact with people face-to-face. One woman we interviewed has participated in online discussion groups for many years. As a stay-at-home-mom in a cold climate, she feels connected to others through the Internet. She told us that she was surprised when she met face-to-face with some of the people she only knew online. "The most hostile people online were the most shy people face-to-face."

The laws protecting people from stalking and online violence have failed to keep up with the problems. A University of Michigan student, for example, wrote four violent stories about another student, using her name and a detailed physical description in the stories. The student posted the original stories to a site with nearly 300,000 subscribers. A University of Michigan alumnus in Moscow saw the stories and notified university officials, who removed the story from the Internet, but the stories continue to circulate among porn-lovers. Lawsuits over the incident were dismissed because the student's postings were protected under the First Amendment (Arnold, 1998). "Harassment online is not the acknowledgment of a woman's sex; rather, like rape, is considered a power game. Men will use words that embody actions to abuse women" (para. 14). Thus, the Web can be a dangerous place when the Internet intensifies problems that go beyond flaming and arguing. The disturbing rape in cyberspace in a user group during the Internet's infancy (e.g., Kaminer, 2001; "Man posts," 1997) has often been cited as evidence of how violent men can become when communicating to women on the Internet.

The violence can seem surreal or like a unique reality, when attacks on individuals which are intensified by the Internet—such as stalking e-mails and harassing websites—threaten individuals or groups of people. The University of Michigan, for example, disabled system software after it contributed to eight incidents of stalking and violence toward women on its campus (Olsen, 2001). Not a new problem, the cloak of Internet anonymity prompts some people to threaten others online.

"Lord, grant me the serenity to accept the things I cannot change ... and the wisdom to hide the bodies of the people I had to kill." "I have a .45." said numerous e-mails sent to a Dallas Internet businessman. In other cases, online chat and discussion have led to real assaults. Laws have not kept up with the Internet harassment, so most people have difficulty stopping the Internet assaults and violators remain free (Whitelaw, 1996). An equally chilling example of violent communication on the Internet is the web site that encourages murdering physicians. "The site's home-page is decorated with dripping blood and doctors who have been murdered have had lines drawn through their names" (Rovner, 1999, para. 2).

"It's actually [sic] obsene [sic] what you can find out about a person on the Internet," a man wrote just days before he shot a woman 11 times. She was outside her workplace; he found her through the Internet (Peterson, 2001). Also among the Internet horror stories are the serial sexual assaults and murders committed by a man who used the Internet to meet women for sadomasochistic sexual rendezvous. The man called himself slavemaster (Rogers & Oder, 2000).

Hate Speech

A father and his son are proud of the 5,000 daily visitors to their web site. At their site, children can find animation, puzzles, games, and songs to enjoy. "My name is Derek. I used to be in public school, it is a shame how many white minds are wasted in that system."

This site designed to teach family values is one of thousands of hate sites on the Internet, specifically, this site is similar to 50 other influential U.S. children's Internet sites designed with the intent to hate, hurt, or murder other people (McKelvey, 2001).

> With only a little searching it is not hard to find sites that are relentlessly anti-Semitic, anti-black, anti-Muslim, anti-Catholic, anti-homosexual or gratuitously violent. Even Sesame Street's Bert and purple dinosaur Barney have a dark side in the hands of some Web site operators. (Sheppard, 1999, para. 3)

*****

Although hatred-inspired violence is hardly a new development (ask Cain), there is a dangerous new ingredient: the Internet. In October, a homosexual group, in the course of investigating an America Online (AOL) policy about "objectionable" speech, discovered scores of postings on AOL advocating violence against homosexuals. Other messages called for blacks to be lynched, Christians killed and Jews burned. The Internet, it appears, gives hatemongers just what they want: a cheap device to reach millions of new recruits. ("Downloading Hate," 1999, para. 3)

*****

The Internet is wonderfully versatile, which is why everyone is turning to it for information and trade. The trouble is, so are criminals. All sorts of crimes are committed using the net, from straightforward hacking to industrial espionage, sabotage, fraud, infringement of copyright, illegal gambling and trade in narcotics, medicines and armaments. The web is also used to peddle child pornography And it is a vehicle for the dissemination of hate literature.

Neo-nazi groups have taken advantage of the Internet to spread their doctrine. Their campaigns, which specifically target young people, encourage racist violence and propagate revisionist lies about the Holocaust. Hateful songs and children's games can be downloaded; one game allows the child to assume the role of a concentration camp commandant. (Sieber, 2001, para. 2–3)

*****

It is a sad commentary that the outrages have become so common that we need a database to keep tabs on who's hating whom. Nonetheless, the experience of the past half dozen years has taught us: Hate groups that burst onto the national scene after acts of terrorism often have names we wouldn't have recognized a day earlier. Further research usually shows such organizations actually have been growing for years and sometimes even have affiliates in our own readership areas, groups we knew nothing about beforehand.

The Internet is frequently an integral part of the hate group, which uses the Web to spread its word and to recruit members nationally, even internationally. (Bowen, 1999, para. 2–4)

Hate speech is a message that advocates oppression and violence toward a group of people. The Anti-Defamation League published its first report "Computerized Networks of Hate" in 1985. Although participation in bulletin boards and listservs was limited at that time, white supremacists provided the forerunner of extremist hate on the Internet. These sites are often designed particularly to entice young people and to take them in through glittering generalities of patriotic themes ("Poisoning," 1999).

Hate sites began on the Internet in the mid-1990s, and their numbers expanded rapidly. Now hate groups in general across the nation are on the rise because of the Internet (Parker, 1999). *Hate sites* advocate violence toward immigrants, Jews, Arabs, gays, abortion providers, and others. Through the Internet, disturbed minds effectively fuel hatred, violence, sexism, racism, and terrorism. Never before has there been such an intensive way for disenfranchised people to gather to reinforce their prejudices and hatred. In one analysis of hate speech sites, the researchers found sophisticated use of persuasive strategies. The hate sites generally started with an objective approach that was straightforward and neutral in which they refuted counterarguments, although the researchers were unable to show that such web sites actually convert people who don't already think their way. The web sites do, however, reinforce and strengthen the hate ideas that people already have (McDonald, 1999). The high school bullies who terrorized kids in the bathroom by writing hate graffiti on the walls and pummeling children have found a new means of harassing and hurting others—the Internet.

People are considered paranoid when they have unrealistic fears about other people being out to get them. People who are the brunt of hatemongering conveyed through the Internet, however, may experience a valid and intensified fear from hatemongers. That is not paranoia because the threat is real. An expression of hate prompts fear among the people toward whom the hostility, anger, and violence are directed.

Consider some other examples. How do you feel—or might you feel—if you were the brunt of this hate via the Internet? The hate group Bully Boys are a pan-Aryan group that seeks the mass killing of Jews, gays, and blacks. The Internet has helped their worldwide network of sponsors for the group's music and their song "Six Million More" (Herbert, 2001). The Internet has provided a vehicle for spreading hate speech of white supremacist, militia extremists, and others. Discussion groups were peppered with anti-Semitic e-mails after Connecticut's Senator Joseph Lieberman, an Orthodox Jew, was selected as the Democratic Vice Presidential candidate for the 2000 election (Johnson, 2000). In an AOL message board following the Lieberman nomination, for example, thousands of comments were posted, including ethnic slurs and Holocaust denials. The same kind of hate-mail was posted on Yahoo (Guernsey, 2000, p. A18). The Internet provides a forum for people with prejudicial attitudes to speak out and act out. Hatemongers can create an online world where they reign supreme, a world of similar minds, where they can gather with others to feel that their way is right and where they can design mayhem for the on-ground world.

Fear of Terrorism

The objective of terrorism is to commit unexpected acts of violence against innocents, which makes other people feel afraid and vulnerable. The kind of small-cell terrorism used by September 11, 2001, terrorists also has been advocated by white supremacists who use the Internet to coordinate their activities. They use the Internet to provide a philosophical framework for "lone wolves," terrorists who work alone or with another individual or two. The murderers are able to work independently under their Internet leadership, and their independence creates anonymity so they are harder to track. "The idea that men who feel they are called by God should commit independent acts of terrorism was put forward by Richard Kelly Hoskins, a former member of the American Nazi Party, in a 1990 book, *Vigilantes of Christendom*" (Thomas, 1999, p. 1).

Perhaps you heard the rumor about Osama bin Laden having hidden secret messages in pornographic web sites. This theory is substantiated by law-enforcement personnel. In addition, terrorists have conducted Internet research and communicated via e-mail. Steganography—Internet hidden messages—"is a fast, cheap, safe way of delivering murderous instructions." The process hides minute images that are too small to be seen by the naked eye. Then the receiver uses special software to decipher the message. The steganography can be detected because expected visual patterns are altered, but with the immense size of the Internet, we do not have the current technology to search for the hidden messages (Cohen, 2001, p. 65).

Today the majority of U.S. residents can imagine themselves or their loved ones being victims of a terrorist attack, and one way they deal with their fears is to discuss their fears online (Kakutani, 2001).

In addition to the elements of creating terrorism through the Internet, there is the threat of attacks on the Internet's infrastructure or attacks on individual business computers through viruses and hacking.

> AN OCT. 6 FBI ADVISORY TO THOUSANDS of law-enforcement agencies and private companies on security threats came as a scarcely needed reminder that the nation no longer has the luxury of complacency. Managers of infrastructure didn't receive that same notice directly, but many already realized the vulnerability of water, transportation, energy and telecommunications facilities since Sept. 11 and are expanding protective efforts. (Armistead, Cho, Ichniowski, Rubin, Angelo, Tuchman, & Sawyer, 2001)

Facing Fears

After the September 11, 2001, attacks, a University of California-Berkeley graduate student, Ka-Ping Yee, set up a database to allow survivors to check on the whereabouts of family and friends. The site had 45,000 valid entries, which served to reassure people seeking information (Berman, 2001).

> How distant we are from that time before E-mail, when chain letters presented us with nothing more difficult to grasp than a claim of being a commercial chocolate chip cookie recipe. (Was yours supposedly from Mrs. Fields? Mine was from Neiman Marcus.) Today, E-mail is still used for recipe swapping, but now we're swapping recipes on how to perform anthrax detoxification of household mail. You favor steam-ironing? Me? The microwave. (Neither technique is safe or effective.) We laypersons are on our own, at least for now. The Centers for Disease Control and Prevention seems to know only one thing: not to repeat whatever it advised us the day before. We'll just have to bluster through with whatever wisdom can be distilled in E-mail folklore. We may not have quality, but at least we've got quantity. (Stross, 2001, p. 44)

We can laugh at the hoax, dispel the rumor, or backtrack information about the stalker. A paradox of the Internet is that despite causing valid reasons for fear, the Internet also enables people to face their fears. The Internet is a powerful source of connection and information for response.

- An e-group of women can discuss sexual harassment at work.
- An e-group of cancer patients can talk about health strategies and coping with death.
- A company can use employee e-mails to provide evidence against them.
- The government can track terrorism through Internet connections.

Reflection

Do you associate the Internet with positive or negative behaviors? Or both? Have you observed polarizing paradoxes emerge in your communication on the Internet (see Table 6.2.)?

The Internet can fuel hatred because individuals can exert their personal power. While feeling safe and in the comfort of their own home, people can use the Internet to speak the unspeakable. We are, of course, talking about Internet use in the United States. What are the implications, for other cultures? Do you think our generalizations hold up to scrutiny in other contexts?

Internet hoaxes can create fear and system problems. By getting people to pass along the hoax information, for example, the message acts like a virus by tying up transmission and server space. Rumors and myths seem particularly credible on the Internet because of their written form and repetition. There is evidence that a normal individual can become addicted to Internet use to the point of neglecting personal and work responsibilities, and becoming socially isolated (Belsare, Gaffney & Black, 1997). Although some people may use the Internet in unhealthy ways, the average user is in no danger of becoming addicted to the computer. The Internet fuels fears and the sense of vulnerability because indeed we are vulnerable. The Internet is used to communicate terrorism, enables stalking, creates opportunity for stealing identities, and facilitates hatemongers. Do you use the Internet to access to information when you are worried? Have you found community and support through Internet Interplay that enables you to confront your fears?

Indeed, polarization of people is a dark side to human communication on the Internet. That dark side includes the hostility of flaming, violent, sexist, and racist language, and the facilitation of murder, terrorism, and genocide. Certainly the Internet uses overt sexism in advertising (Artz, Munger, & Purdy, 1999) and in the sexual content on the web, which uses sexual language (Noonan, 1998). In fact, in a comparison of Internet, video, and magazine pornography, one study found significant differences between Internet pornography and other media. The pornography on Usenet (Internet news group) was significantly *more violent,* more coercive, more nonconsensual, and more often represented men as dominating and victimizing women than the pornography in other media (Barron & Kimmel, 2000).

TABLE 6.2 *Polarization Paradoxes*

Make people feel safe because they can find out what they need to know.	Increase fear by fueling hatred. Fuels fears and sense of vulnerability.
The violence doesn't feel real.	Flaming, violent, sexist, and racist language, and use of the Internet as a facilitator of murder, terrorism, and genocide.
Online feminist and activist groups.	Sexism in advertising.
Easy access to information.	Lack of privacy.
Solid information.	Hoaxes, scams, viruses.

Case for Discussion

In your course discussion group, give your analysis and perceptions of the following.

Natalie is a U.S. citizen, born in Hungary, who lived in Israel for a while. Natalie noticed that the widespread use of cell phones was greater than in the United States. At first she thought it was a cultural aspect of the way people communicate. After living for a while in Israel, a place where a terrorist might strike at any moment, anywhere, Natalie realized that the cell phone was neither a business nor a cultural communication device. Natalie says the cell phone is a vehicle for caring.

As a way to engage each other and keep track of the whereabouts of loved ones, people frequently call each other to check in, hear each other's voice, to say "I love you" one more time, and to reaffirm that all is well. During a terrorist attack, the cell phone enables people to connect so they know loved ones are safe. And if the unthinkable happens, they can say "good-bye."

As technology converges the cell phone, wireless computing, and the Internet, our human communication options increase. Right now, when you use your home computer to connect to your boss at work, you connect from a place to a place. Wireless Internet allows you to connect from person to person.

Internet Investigation

1. *Archive Investigation.* Explore archives of an e-mail or messageboard e-discussion group. Find an example of hostility or flaming. Do you think the same words would be said face-to-face? If not, rewrite the interaction to read the way you think two people would speak if they were face-to-face.
2. *Gossip.* An angry person online can spread gossip and lies directed at an individual. Such a nightmare happened to a Russian scientist who worked for the National Institute of Health in the United States. A comment the scientist made about the Holocaust was misquoted and posted in a vicious note to hundreds of newsgroups, including ones on Jewish topics. Although his employer said he was not at fault, they took away his online account, required he submit to psychiatric counseling, and terminated his work visa (Schwartz, 2001). What recourse do you have when someone spreads rumors and lies online? Why are online lies more dangerous than ones circulated face-to-face? How can you respond online, particularly in the work context, that will not come back to haunt you?
3. *Debunking Urban Legends.* There are many sites designed to counter the rumors, gossip, and lies on the Internet. Investigate, explore, and write an essay about what you discover.

    Among the respected sites designed to debunk urban legends are:

    ScamBusters at http://www.scambusters.org
    The Urban Legend Combat Kit at http://www.netsquirrel.com/combatkit

www.fsecure.com/virus-info/hoax
www.symantec.com/avcenter/hoax.html
www.snopes2.com rates current rumors

4. *Hate Watch Group.* Find one of the hate watch groups online. What do they say about the current state of hatemongering and terrorism?
5. *Digital Divide.* The Markle Foundation exists to improve the quality of life in the information age. One of their projects is to study the "digital divide" caused by online haves and have-nots. Examine the work of this foundation and write a 200-word essay that discusses the implications for health communication [http://www.markle.org/index.htm].
6. *Cultural Influences Discussion (Hate).* We have discussed only certain types of hatemongering on the Internet. There are men and women of various ethnic, religious, and cultural affiliations who spew hate through the Internet. Investigate the Internet hate group from a different perspective than one discussed in this chapter and post your findings to your online discussion group.
7. *Mass Communication View of Persuasive Technologies.* As you ponder the ideas of this chapter, how might you see the ideas through the eyes of an expert in mass communication? Investigate the concept of captology or the study of persuasive technologies of computer interaction. There are all kinds of software available, which are designed to change the attitudes, beliefs, and behavior of people who use that technology. The persuasive nature of mass communication is a significant area of scholarly research. Investigate the persuasive nature of the Internet and computer use. What is the persuasive nature of Internet design? Layout? Delivery? For information on captology, see Fogg, B. J. (1999). Persuasive technologies. *Communications of the ACM, 42*(5), 26–29.

Concepts for Analysis

We have provided concepts relevant to this chapter so that you can check your understanding. These concepts are worth considering in a more critical way. As a research option, examine one concept. Use our citations as a starting point for your study. We recommend that you search for refereed articles in an online database. Find at least three solid references you can read. Summarize, scrutinize, and present your analysis to other people in your course.

Questions to stimulate your inquiry: Do you agree with the ideas as we presented them? What other points of view should be considered? What details need to be added to fully understand the concept? What can you contribute through your personal analysis of the concept? What more do scholars need to know about the topic?

Cognitive dissonance theory suggests that our beliefs or cognitions can be related either by their consistency (consonance), their contradiction (dissonance), or their irrelevance (lack of connection). In general terms, we tend to hear and seek information that is consistent with our beliefs. We tend to avoid or refuse to believe information that is inconsistent with our beliefs (dissonant) (e.g., Brehm & Cohen, 1962; Festinger, 1957; Petty, Ostrom, & Brock, 1981).

The *democratization* ability of the Internet is the idea that information flow and communication access empowers United States citizens so they have greater opportunity and equity.

The *digital divide* is the idea that the Internet creates more inequality because only the advantaged can have and use the Internet effectively.

A *discussion group* is a gathering of people who interact together via the Internet. They often have a common bond of some kind, such as that they like to discuss politics, sex, or a hobby. The group may use e-mail discussion to interact. Often the e-group uses a discussion board or webboard, which is a service set up so that electronic communication is conducted on a server.

Flaming is when people exchange hostile or insulting remarks.

Hate speech is a message that advocates oppression and violence toward a group of people. Hate sites advocate violence toward immigrants, Jews, Arabs, gays, abortion providers, and other people based on their culture, ethnicity, or beliefs.

Internet addiction is a condition in which people cannot control their behaviors, so they use the Internet to the extent that the Internet interferes with normal activities

An *internet hoax* is an elaborate practical joke that conveys false information through the Internet.

Metaphors are words that create meaning by using words that are an analogy or a comparison. When we call the Internet the "Web," for example, the metaphor brings to mind the reach of a spider web.

Netspeak is a new language dimension, that is, computer-mediated language is unique through its elaboration and enrichment of conventional language (Crystal, 2001).

Polarization is the ability of the Internet to encourage groups of people to emphasize their differences with others so they are repelled by other people.

Portals are entries to comprehensive Internet services, including search engines, category guides, evaluations of the credibility of other information sites, and services, that can be quite easy for the novice to use. The objective of these portals is to make the Internet easier to use so that people can access the information they need.

Terrorism is the use of unexpected acts of violence against innocents with the purpose of making other people feel afraid and vulnerable.

Urban legends are a type of hoax that spreads myths. Perhaps Internet hoaxes are designed to be an extension of the human desire for humor and an appreciation for practical jokes.

Part III

Communication Modes

This section explores communication on the Internet through the perspectives of three communication modes; intrapersonal, interpersonal, and group communication. We stopped short at public speaking and mass communication because we believe communication on the Internet to be primarily a conversing or interpersonal mode. We do, however, believe that public and mass communication frequently occur on the Internet, although with the Internet, any kind of communication can become public!

Computer-mediated public communication includes teleconferencing, online meetings, public archives of e-groups, and web sites. But we find that the Internet blurs modes of communication. In fact, the various modes converge into a new mode of communication: Internet-mediated communication.

Chapter 7—Intrapersonal Communication as Cognitive Collaboration—Gives you an explanation of how people come together mentally through their online conversing.

Chapter 8—Interpersonal Communication On the Internet—Gives insight into how people build and maintain relationships online.

Chapter 9—Groups—Gives perspective on how communities emerge online.

7

Intrapersonal Communication as Cognitive Collaboration

> FORGET THE MYTH ABOUT ENGINEERS WORKING ALONE IN A CORNER *solving major design problems. It was never completely accurate. Engineers have always found ways to brainstorm, whether over the phone or by passing sketches back and forth on cocktail napkins.*
>
> *But Internet technology has increased the number of eyeballs looking at those napkin sketches—and made it possible for teams scattered across the globe to comment on designs as if they were all in the same conference room at the same time.*
>
> *Indeed, the Internet has helped make collaboration one of the biggest trends in design engineering today. And, its ability to bring teams together holds the promise, some say, for a major improvement in the quality of designs.*
>
> —(Porter, 2001, para. 2–4)

Collaboration is communication that brings people together in their thinking or spirit. When you communicate intrapersonally, you are communicating with your self. "Is that the same as thinking?" you might wonder. No, intrapersonal communication is a mental "dialogue" that implies a reaction to your self or within your self, a reaction to an imaginary person, or an imagined other. *Cognitive collaboration* is when two people share intrapersonal processing or jointly construct ideas, so that they become of similar minds. We argue that Internet interplay intensifies that immersion into others by creating a sense of intellectual communication or cognitive collaboration.

Focus

```
TO: Students

FROM: paradoxdoc@hotmail.com

RE: Overview: Intrapersonal
```

Lenny and Joan consider themselves experts on intrapersonal communication processing. So this section offers their particular viewpoint about how the Internet is a way for people to collaborate mentally. Collaboration is communication that brings people together in their thinking or spirit. Cognitive collaboration is when two people share intrapersonal processing or jointly construct ideas, so that they become of similar minds. Lenny and Joan argue that Internet interplay intensifies that immersion into other people by creating a sense of intellectual communication or cognitive collaboration.

Lenny and Joan talk about intrapersonal communication as assigning meaning to stimuli and producing meaningful stimuli, regardless of whether those stimuli are verbal or nonverbal. Their fascination with computer-mediated communication—specifically Internet-mediated communication—comes from the belief that the computer is an intensifier of the intrapersonal aspects of communication. The Internet increases a person's fascination with the self as the computer is an extension of the self.

Lenny and Joan chat about inner speech, outer speech, the mediated symbolic process, and the inner/outer dichotomy. Lenny and Joan mention their cognitive collaboration on the Internet while writing this book, for example, and they discuss other ways people collaborate cognitively.

A key point to understanding their discussion is the theory that online interplay affects how individuals process their communication. Lenny and Joan ask you to consider the relationship between the self and the Internet. And they discuss how some people attribute human characteristics to computers. Other variables for you to think about relative to cognitive collaboration

> are role-playing, culture, age, and gender. The Internet
> creates cognitive and social interaction, and the
> interplay of that cognitive and social interaction can
> intensify mental or cognitive collaboration.

This chapter will explore Internet paradoxes of intrapersonal communication processing that enter into cognitive collaboration. At the end of this chapter, you should be able to answer the following questions.

1. How does cognitive collaboration resolve the inner/outer dichotomy?
2. What is the link between the self and the Internet?
3. How does the Internet create interaction?
4. How does role-playing intensify creativity about identity?
5. How might U.S. cultural values fit cognitively with Internet use?
6. Is the Internet a positive or a negative for the young and elderly?
7. Are there differences in how women and men perceive communication on the Internet?
8. What are the paradoxes of cognitive collaboration on the Internet?

Intrapersonal Communication and the Internet

The framework for our work is twofold: paradox and meaning. We have introduced the idea of paradox and we opened the idea of assigning meaning to messages sent over the Internet. We already suggested that intrapersonal communication is involved in the meaning-making process. It may seem odd to some people to talk about what goes on inside an individual's mind when speculating about communication on the Internet. Paradoxical perhaps, but intrapersonal communication seems to fit. While we are looking out at the larger world through the Internet, we find ourselves simultaneously looking in to our minds and our assigning of meaning to symbols. Perhaps the most efficient way to introduce you to the idea of intrapersonal communication and its relationship to the Internet is to offer a model (Figure 7.1).

If you think of the intrapersonal model as representing you as an individual, then picture the stimuli coming into your receptors (senses of hearing, sight, and so on) as coming from the Internet. Also picture the stimuli you send via the Internet as expressed through your *effectors* (hands, speech, and so on), which are mediated by the computer.

When all is said and done, the reason for communicating with one another is to create meaning, to perform social actions. Your memory system is built to recognize that the meanings are what is crucial, what is to be stored away. In addition, there are many kinds of meaning—or levels of understanding—not the least of which is how you feel in connection with a communication event. In most situations, you probably remember *how you felt* about a communication *more than what was said*. While the interaction of the inner and the outer are melded together, the outer is shed as you attach meaning and response.

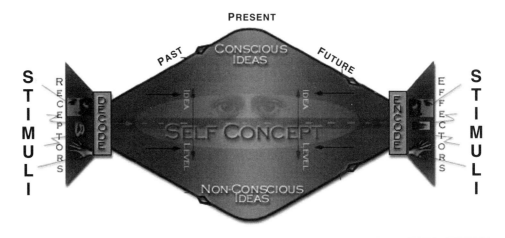

FIGURE 7.1 *Intrapersonal Communication Model*

Perhaps this is the point at which information and meaning depart down significantly different roads. We can think of information as one level or one type of meaning. Think of this level as akin to literal meaning or propositional meaning. The Internet engages us in a playful world of work, learning, entertainment, play, romance, distraction, pleasure, friendship, security, and connection with human intelligence. The process by which this happens is human communication. Human communication relies on multiple levels or types of meaning.

Intrapersonal communication is communication to the self. We define *intrapersonal communication* as assigning meaning to stimuli and producing meaningful stimuli, regardless of whether those stimuli are verbal or nonverbal. All intrapersonal communication uses cognition—thinking, reflection, or mental processing—but not all cognition is intrapersonal communication (Winograd & Flores, 1987). Intrapersonal communication requires a type of self-awareness, a recognition of the process of communication. In intrapersonal communication, we don't really care whether stimuli are generated within the perceiver (i.e., the intrapersonal communicator) or are received from outside the perceiver. Nor does it matter to our definition of intrapersonal communication whether or not what is generated remains within the individual (e.g., talking with one's self, "seeing" images, or having physiological sensations) or somehow ends up outside the person (e.g., a journal as a computer file). Sometimes what is generated as intrapersonal communication is not expressed or made observable to others (self-talk), and sometimes what is generated is expressed (e.g., self-talk overheard by another person, an online journal).

> The study of intrapersonal communication is also, therefore, the study of decoding and encoding—the study of meaning-making. Whether or not what is decoded originates inside or outside the body of the intrapersonal communicator, and whether or not what is encoded is actually expressed, leaked, or given off, intrapersonal communication has occurred (see Goffman, 1959). Stated broadly, intrapersonal communication is about the relationship between the individual and the stimuli that the individual encounters. (Shedletsky, 1989, p. 94)

You might wonder if intrapersonal communication is the same as thinking. No, intrapersonal communication involves both internal and external forces. When you "talk" to yourself aloud, when you use language to process experience, when you imagine a conversation you may have with another person, then you are communicating intrapersonally. On first thought, *intra*personal communication and the World Wide Web would seem to be far apart from one another (see http://www.usm.maine.edu/~com/intramod.htm). Intrapersonal communication points to the mind of an individual; computer- and Internet-mediated communication prompt images of connections between millions of people around the globe.

Our fascination with computer-mediated communication, specifically Internet-mediated communication, comes from the belief that the computer is an intensifier of the intrapersonal aspects of communication (Shedletsky, 1993). As professors, we have observed that working online intensifies student understandings of intrapersonal communication, while study of intrapersonal communication intensifies student understandings of computer-mediated-communication. Intrapersonal and Internet communication reinforce one another synergistically (Shedletsky & Aitken, 2002). This idea has prompted the creation of this book as a more comprehensive analysis of human communication on the Internet.

The study of intrapersonal communication is *metacognition.* And the World Wide Web component reinforces the social nature of our cognitive selves (Martin, 1997).

> Human cognition is not an island unto itself. As a species, we are not Leibnizian Monads independently engaging in clear, Cartesian thinking. *Our minds interact* [emphasis added]. That's surely why our species has language. And that interactivity probably constrains both what and how we think (Harnad, 1995, p. 397)

In other words, the inner world has developed as it has in response to the outer world—the cognitive and the social are inextricably woven together. Language, thought, and social interaction are packaged together. Communicating over the Internet helps users to see that combination more clearly.

The tension between inner and outer—between mental and material—represents the intersection of philosophy, science, and social science. As scholars have adjusted their thinking about mind and behavior, they have expanded their thinking to accommodate both the inner and the outer worlds. New concepts and new ways of behaving require new understandings. The Internet is one of those new concepts and new ways of behaving that require us to reshape our understanding. We see this struggle over coming to grips with the new ideas in the proliferation of metaphors to talk about the Internet.

> A measure of the unsettled conceptual state of the online world is reflected in the wide range of metaphors used to talk about computers, in the range of emotions we express about computers, and in the ways we commonly use computers. We speak of them as servants, tools, work areas, toys, filing systems, fancy typewriters, and global spaces. We speak of computers as scientific instruments and as devices of both interpersonal and mass communication. In the 1950s we thought of the computer as a very fast calculator, able to do arithmetic problems more quickly than any human. We used the word "cybernetics" to refer to

computerized messages controlling machine actions, as we could expect to see in a robot. These were functions cold and removed from ordinary human communication. Today, we fill up our personal computers with information of all sorts; we use our computers to travel through space and time; to shop; be entertained; acquire information about the latest movies or today's headlines; talk to colleagues—some of whom we've never met in person; download music; do research; meet with friends; send and receive documents and pictures; send birthday cards, father's day cards, and such. We surf on our computers or take highways; we work on our computers, play on them, play with them, play against them; we invest in the stock market; we both create and visit communities; we interact with both the hardware and the software; we see pictures there, read and write words, hear sounds. People love their computers, hate their computers, are addicted to their computers, feel they are discriminated against by their computers, and are frustrated by their computers. The computer takes us deeply into our own thoughts and off into the thoughts of others. (Shedletsky, 2000, pp. 164–165)

Inner/Outer Speech

Inner speech is communication to yourself in which you talk to yourself internally. Inner speech is when you use words to sort information and attach meaning to what is happening. How does inner speech compare to interpersonal communication with an intimate friend? Inner speech is more abbreviated and rapid. *Outer speech*—when you speak to another—needs to take into account what the other person knows and what they don't know. To make sense of communication with another, you need to fill in information or, in some instances, not state information that is irrelevant or given. If you write for the public, you will be even further removed from inner speech than when you have an ordinary conversation. In your public communication—such as in a letter to an editor—you do not receive immediate feedback from your interlocutor and may not even know who is reading your words.

Now consider your communication on the Internet. The vast majority of people use the Internet primarily to communicate with family and friends (Watt & White, 1999). Granted there are new technologies that enable people to interact through sound and video on the Internet, but the primary use of the Internet is still written, text-based e-mail. When you write an e-mail to someone on the Internet, you assume the person will read your words, but you do not receive immediate feedback unless you happen to be online at the same time and can engage in simultaneous communication. You cannot adjust your messages as you do in face-to-face conversation. When you read another's e-mail message, you cannot hear the person's voice or see their face or gestures or experience the rhythm of speech. Because you are further removed from face-to-face conversation you are required to think hard about the communication event. You have to analyze, scrutinize, ponder. You are driven further inside yourself to your inner speech because you must figure out the communication event so you can attach meaning. You probably ask yourself questions and think through the meaning of what was written. In this way, e-mail communication on the Internet is an intensified cognitive event so that your inside and outside engage (for discussion about this general concept, see Dale, 1976).

The Internet encourages you to reflect on the process of assigning meaning. At the same time, the Internet promotes active learning and the ability to work cooperatively in teams with people who think differently from oneself. In short, we believe that in order to understand human communication on the Internet we must take into account both stimuli produced and sent via computer and the intrapersonal processes involved in making sense of those stimuli. We believe that there is a strong, although complex, relationship between these two domains, the intrapersonal and the computer-mediated (see Campbell & Neer, 2001; Gackenbach, 1999).

Thus, part of the framework of this book is an intrapersonal model of communication. According to this intrapersonal model, the internal versus external tension is woven together as a seamless interplay, a constant "dialogue" between the two. Early models of human communication tended to be linear, moving from point A → to point B → to point C, and so on. Those models were later rejected and replaced with circular models, which were less concerned with starting points and ending points. Those models no longer perceived information as a thing. With the adoption of Berlo's model (1960), for example, we came to be more concerned about the active process of human communication and interpretation of meaning. Consistent with Berlo's explanation of communication as a continuous and changing process, so too is CMC and Internet communication more of a process than a thing. The Internet is changing, evolving, free of boundaries, and instantaneously expanding. Thus, we adopt a model of human communication that is highly interactive (see Figure 7.1, also online at: http://www.usm.maine.edu/com/intramodel/intrapersonal/).

With this intrapersonal perspective, however, the focus is still on social action. With an emphasis on social action, the intrapersonal model is closely associated with the interpretive perspective. Those of you familiar with communication theory may think of concepts such as Social Construction of Reality, Symbolic Interactionism, Frame Analysis, the Dramatism of Burke, or the Coordinated Management of Meaning (see e.g., Infante, Rancer, & Womack, 1996; Littlejohn, 2002; Wood, 1999). If so, you are tracking along with us perfectly. If not, in the pages of this book we make connections to some communication theories that we believe offer insight into analyzing and understanding Internet communication.

For most of us, communicating is largely an experience of dealing in a transparent medium—we have difficulty noticing how language and cultural context mediate our communication. Most people find one-to-one communication fairly easy, natural, and don't think much about what they are doing. Communicating on the Internet, however, makes it more difficult to see through the medium. While communicating via the Internet, we are apt to notice more about the processes of communication and how our minds and hearts interplay with other people. We present symbols to one another. We interpret the meaning of symbols. We respond to our interpretations. The central fact of our experience on the Internet is mediation. The Internet intensifies our awareness of the communication process. Vygotsky would say that the central fact of our psychological behavior is *mediation* (1982). Langer would tell us that the central idea of our time is *symbolism* (1962). We think of communication on the Internet as a *mediated symbolic process.*

The Internet offers a new intensified interplay between the cognitive and the social. It may seem ironic at first, but the Internet intensifies and highlights the mind and its workings. And much like other process models of communication, the intrapersonal model (see Figure 7.1) directs our attention to mechanisms of the mind that are involved in meaning-making.

Resolving the Inner/Outer Dichotomy

Is true human communication possible? Is there any way to convey an idea so the other person thinks and feels exactly the same way? Most communication scholars would say "probably not." Yet we all have times when we believe that we achieve a surprising and rewarding depth of understanding with another person. You know what it feels like to work with another person and be of like mind with that person. In earlier chapters, we discussed how, as a means of communication, the Internet is an intensifier of the intrapersonal aspects of communication (Shedletsky, 1993). Cognitive collaboration is the interplay of the inner and the outer, your intrapersonal and interpersonal communication. Understanding symbolism and interaction are key elements in understanding how you use the Internet. In this chapter, you will explore the cognitive or mental aspects of the Internet and how people use them to interact, come to common meaning, and understand others on a cognitive or intellectual level.

The human action perspective (see Winch, 1958) serves as one viable way of analyzing human communication on the Internet because of the subjectivity of the computer communication experience. In many ways, the important element is how the individual experiences communication on the Internet. To understand the nature of the subjective experience, consider how individuals vary in their understanding of effective communication on the Internet. How does virtual reality create concrete reality for each individual? What is the nature of cognitive collaboration on the Internet? A person's physical and cultural differences are easily hidden on the Internet in ways that can increase one's receptivity to others. One cannot observe behavior or nonverbal reactions, so each individual easily imagines and interprets the other in his or her own way.

There are some essential features of communication over the Internet, which play a critical role in explaining how people collaborate cognitively through the Internet:

> *Mediation* is consideration or negotiation which takes place on the Internet. Similarities between people can be created in the minds of each individual because Internet users seek commonality, which they can find in their intellect.
>
> *Communication* is reaching a common understanding through interplay.
>
> *Cultural mediational artifacts* are the nonverbal communication and cultural information that help people interpret communication. The physical and cultural differences are easily hidden in communication on the Internet.
>
> *Context* is the situation, environment, or situated meaning. The Internet alters and reduces the relevance of the physical context. In a sense, the Internet transcends the time and place of communication.
>
> *Mind* refers to the cognitive processes of thinking. The Internet intensifies the role of the mind in attaching meaning.

Much of e-discussion takes the form of thinking about something, pondering. Except for applications like Instant Messaging which moves very quickly, typically we have more time online than in face-to-face communication to think before deciding on what something means or before "speaking." We get time to edit. Of course, we can also "speak" and send messages online without carefully planning. That often happens in instant messaging or chat or even e-mail—it is as if we are thinking out loud in those cases. In the edit mode, with the added consciousness of the minding process (or meaning-making or interpreting), we are more inclined to be aware of how we figure out meanings. We may even stop to think of how this might "sound" to our receiver. The nature of online discourse has unique characteristics (Brewer, 1998; Ganesh, 2000).

Think about how your mind works. You are not entirely language-based when you are thinking, but you receive flashes of images, memories, phrases, sensations, a montage of incomplete thoughts. When you verbalize those thoughts by talking or writing, you give structure to your thinking. If you write an essay for a course assignment, for example, you may write several drafts as you formulate your ideas. In fact your own thoughts may not become clear to you until you have written and rewritten them. So too is the nature of cognitive collaboration on the Internet (Aitken & Shedletsky, 1998). In short, it may be that both more time to think about the meaning of what we say and receive (encoding and decoding modes) and more opportunity to think on the screen (e.g., reflective processing time) may intensify intrapersonal communication.

We were engaged in cognitive collaboration on the Internet while writing this book, for example. One of us wrote, then the other added, subtracted, and reshaped the ideas. The material went back and forth between us through the Internet as we tried to formulate our thoughts. We have read about, thought about, researched about, and spoken to each other about communication on the Internet. But our ideas continued to change and interplay with each other as we refined, rewrote, and polished. In a sense, when we verbalized and responded to each other, we helped figure out what we thought about these topics. We never used the telephone and never worked face-to-face during this time. We sent e-mails and attachments back and forth. We talked online about the book almost daily, and we were just as likely to write a pun or share a story about a family member at any given moment as we were to write an idea about communication on the Internet. We stopped writing because we reached a time deadline of the publisher, not because we had figured everything out about the subject of communication on the Internet. At the point when we stopped our interplay—no longer exchanged ideas back and forth—our cognitive collaboration ended. And what is the point of this example? To show how people can use cognitive collaboration to arrive at meaning through their communication on the Internet. This mental interplay is part of what the Internet offers. Each of us communicates with the Self and with others as a way to attach meaning to our experience. And that cognitive collaboration is the focus of much Internet interplay.

Think about how you communicate on the Internet. Imagine that you are in a chat room talking about something you heard in current world events. In the chat room you express your opinion as do others. You may present facts and arguments, as do others. The interplay between people affects what each person thinks and the meaning each attaches to her or his understanding. The interaction doesn't necessarily persuade you to think a certain way. You might leave the chat room more entrenched in your previous beliefs. You might leave the exchange totally confused. *But the online interplay affects how you process the communication.*

Now consider how people write to one another online. Some people write carefully, then read and revise the messages, even using spell check. Except in business contexts, however, most people write quickly, don't reread their message, and don't bother with their spell check. Many people write without punctuation, use emoticons, and ramble in an oral communication style. A message might look like this.

come on—you don't really think that will happen LOL—figure . . .

If you were to go back and look at your instant message exchange a week after you wrote it, you might not be able to figure out the meaning of the conversation because out of context it may seem like a bunch of jumbled thoughts. On the Internet, you may write as you think, in a spontaneous stream of consciousness. This is similar to a face-to-face conversation. You attach words to your cognition because sometimes online communication is simply expressive communication. The communication might be simply to vent or express feelings. Or the communication may be for interactive rather than informative reasons. The people may just want to connect or create a type of an online handshake or online hug. Or the communication may be an attempt to make sense of internal processing and be a type of externalization of intrapersonal communication.

What else happens in the process of communicating via the Internet? The Internet enhances the information processing aspects of communication, which enables the assignment of meaning, the use of implication and inference, and the determination of social action. Internet-mediated communication encourages the mind to work even harder than in other modes with regard to focusing upon *voice* (your point of view), *tone* (mode or expression), and *intention* (aim or purpose). Internet-mediated communication brings to consciousness fundamental aspects of communication that are more often transparent in other modes, such as *turn-taking* (alternating messages from one person to the other), *juxtaposition of utterances* (matching what one person says to the other), and *ambiguity* (lack of predictability). Internet-mediated communication promotes cognitive reorganization.

In most face-to-face conversations, people take turns. They stay on topic and are quiet when the other person is speaking, or they encourage the other to continue. These standard rules of interaction break down on the Internet. In Internet-mediated communication, message coherence is disrupted, overlaps and other breakdowns in turn order result in failure to stay focused on the topic. Research suggests that these rules are less important than in face-to-face communication because of the nature of the interaction and language play on the Internet (Herring, 1999).

There are moments when communication on the Internet feels totally simultaneous, enabling two minds to function as one. The two minds working together become more than either person might be individually—a case of the whole being more than the sum of its parts. That kind of cognitive collaboration can be a unique interplay. Remember, the interaction generally happens through written exchange. Consider the implication of this idea for written Internet communication: "The change from maximally compact inner speech to maximally detailed written speech requires what might be called deliberate semantics—deliberate structuring of the web of meaning" (Vygotsky, 1962, p. 100). The Internet offers a new blend of media, and with that, a new blend of dimensions along which humans create symbolic realities. Over the Internet, you generally write to specific individuals. You do not have access to the person's immediate outward reactions to your message. You cannot be certain you are "heard" and understood. The time delay between writing and receiving the e-word is not as long as with a letter to the editor, however. Your cognitive, inner processes and knowledge, your outer messages, and the communication of meaning are all affected by this new way of sending and receiving messages.

The Self and the Internet

Consider the case of a high school senior male who used the Internet more than 30 hours per week. After observing pornography and her son's other Internet activity, the mother wondered if she should expect a Federal Bureau of Investigation agent to show up at her door some day. The teenager's skills were unusual, as demonstrated by the fact that a large American communications company asked him to write a program they planned to use. In describing the main effect the computer had on him, the teenager said computers had given him "a name, that's what I'm known for." The teen perceived the computer as his source of identity. Although an atypical example, his perception shows how the Internet can shape the self.

While using a computer, an individual may be communicating primarily with the self, but that communication is mediated by the way the computer program works, the information on the Internet, even posting to a discussion group. The posts to a discussion group may never be read by anyone else and never receive a response except internally (by the other), yet the sender has a sense of speaking to another and by perceiving the communication as a form of intrapersonal communication (for the other), the sender has a sense of collaboration with someone else. And of course all the while this is an intrapersonal communication experience for the sender. The experience is rather like the child who writes in a diary or journal imagining that someday it will be published as a famous life story. So too, the Internet discussant may be writing to the self while imaging a reader. The result is mediated-intrapersonal communication. Sometimes the individual doesn't know whether she or he is interacting with someone else online. And at other times, another person becomes clearly involved in the musings of the individual, sometimes musings that were intended as a type of self-communication. There are, in fact, discussion groups designed to be online journals of self-expression and self-exploration, in which the participants are required to post a journal entry once a week. Those postings may, however, be read by other individuals who give feedback to what is intended to be intrapersonal communication, making that intrapersonal communication a collaborative experience.

The computer serves as the *second self* by engaging the individual in intrapersonal communication and self-exploration available in no other way. Consider this quote from a classic work that formulated the idea.

> Computers call up strong feelings, even for those who are not in direct contact with them. People sense the presence of something new and exciting. But they fear the machines as powerful and threatening. They read newspapers that speak of "computer widows" and warn of "computer addiction." Parents are torn about their children's involvement not only with computers but with the machines' little brothers and sisters, the new generation of

electronic toys. The toys hold the attention of children who never before sat quietly, even in front of a television screen. (Turkle, 1984, p. 13)

Among her many interests, Turkle (1984) noticed the role of computer jargon in vocabulary, a vocabulary that includes "psychological discourse." The similarities between computers and people arouse a unique vocabulary, as she explained: "Their language carries an implicit psychology that equates the processes that take place in people to those that take place in machines" (pp. 16–17). The implications of her words are astounding: People are using computers and the Internet to understand people. Some people can understand certain characteristics of the computer better than they can understand humans because the computer is simpler and because the computer and human system are similar in some ways.

Research has suggested, for example, that a person's e-mail use demonstrates characteristics and reflections of the user (Tao, 1995). Although some research has indicated that e-mail users appear self-absorbed—probably because e-mail lacks normal social cues—other research has suggested that users are not perceived as self-absorbed (Adkins, 1994). We propose that the people in e-mail discussion groups are no more self-centered than anyone else, but characteristics of the computer-mediated communication intensifies expression of one's self-absorption, that is, one is free to express opinions without responding directly to others who also express their opinions. We see the self-absorption as positive because we hypothesize an emphasis on self-awareness and sharing.

The connection to self, prompted by the networked computer, can make online discussion an appropriate vehicle for learning about intrapersonal communication, for example, because the networked computer is an intensifier of the intrapersonal aspects of communication (Shedletsky, 1993). In other words, the computer enhances or brings awareness of the information processing aspects of communication, the assignment of meaning, and the use of implication and inference. CMC encourages the mind to contemplate the message, the *metamessage*—or larger, holistic meaning of a message—and the social action intended (e.g., making a promise or offering something). CMC brings to consciousness fundamental aspects of communication that are more obvious than in other modes. While "talking" on the Internet, you are inclined to consider communication aspects such as taking turns in a conversation, the order of what people say, placement of messages (Nofsinger, 1999), context, confusion over meaning, silence and play on words. The computer promotes cognitive reorganization, reflection on the process of assigning meaning to messages, and, at the same time, promotes active learning and the ability to work collaboratively in teams with people who think differently from oneself.

If you are an avid Internet user, for example, you probably have felt surprised when you suddenly realized how much time had passed after you had intently focused on the Internet for awhile. Consider the concentration levels exhibited by one couple we interviewed. One man quit his management job, decided to write a book, and became totally engrossed with a computer in the project. According to his partner, he "spent a year in the bedroom on the computer . . . there were times when [he] would disappear for days." The husband kept a "do not disturb" sign on the door. When the wife took a computer programming course, she understood the involvement of the computer: "I can sit for 10 straight hours" she explained.

Perhaps you felt a sudden rush of physical awareness after intense Internet use. You abruptly felt awareness of your physical reality when you had been operating in virtual reality. We do not make the distinction of reality versus unreality because physical reality and virtual reality or "onground" reality versus online reality are both quite real. But that sense of internal play gives the Internet user the impression that something unreal or surreal is happening because most people have an engrossing ability for intrapersonal processing while on the Internet. If you have ever tried to pull someone's attention away from the computer, you understand. The Internet intensified interplay can be similar to meditation, where one's mental focus allows one to forget about physical surroundings.

We have used the Internet to teach intrapersonal and interpersonal communication courses because of this ability of computer interaction to enable cognitive collaboration. There is a sense of openness and mental connection that just doesn't seem to happen in our traditional classrooms. By way of example, consider the format used for teaching an interpersonal communication course.

The course had a mix of women and men, different U.S. ethnicities, and some international students. Each student had to write a researched essay each week on one of the topics of a case study. That essay went to the teacher and the feedback was directly to the student. In addition, students were assigned to read two or three case studies each week which they discussed online. In WebCT course environment, there was an online messageboard for each case. The students had to discuss at least one case each week. Student discussion grade was based on the number of postings they read and posted. The instructor responded through private e-mail regarding poor quality content of messages. Twenty-five postings were required for a C grade. The instructor didn't keep track of how many postings were available for students to read. She provided a bunch of research findings and definition of terms/topics. Most students posted 30–60 messages online. Some students read more than 1,000 postings, most read about 200, many read about 600. The instructor was stunned by the frankness of some discussions. Here are some examples:

> Student 1: When you think of interpersonal communication, the last thing you think of is the Internet. But, because we did case discussions online it did become interpersonal. It was so interesting reading everyone's ideas on our society and culture and the families they grew up in. We all agreed and we all disagreed. We had nothing and everything in common. I am glad that we did these case discussions over the Internet because there are many things that we all shared in writing that I KNOW we would have never been able to share in the classroom. For many people, it is easier to put what they feel or say in writing rather than speaking it. Because some of the topics were so personal and hit home for many of us, communicating online really enabled us to share a part of ourselves.

*****

> Student 2: We discussed much delicate subject matter and frankly, I don't know that I would be THAT willing to share amongst a discussion full of mainly strangers! In reading the posted comments I discovered the trust in the uniqueness and individuality of people and how each person's life experiences shape how they think and feel. Definitely eye-opening.
>
> I am not the type to speak up in class but the way this class was organized, it made it a lot easier for me to voice my opinion, and if I didn't my grade would suffer immensely! I felt much more comfortable about speaking up when it was done over the Internet, in sort of an anonymous manner.

*****

> Student 3: The class did not meet twice a week in another room with a small percentage of students skipping every week and several students daydreaming until it was time to leave. Instead the class got to post comments on the Internet and actually interact with one another. On most weeks, every student in the class participated and posted multiple comments. I felt that this was the best way to allow everyone to voice his or her opinions on many difficult topics. It was also a good way to let normally shy people get involved in class discussion.

Other students discussed how posting messages and interacting on the message boards seemed like writing in a journal or diary. They compared the Internet public discussion to private communication to the self (Aitken & Shedletsky, 1998).

Attributing Human Characteristics to Computers

The process of attributing human characteristics to a computer may seem strange at first. Yet people have long given such credit to inanimate objects, such as trains, favorite cars, and ships (Ingber, 1981, p. 91). The difference with the computer may be that it really does have many human features, so computer personification by some users appears somewhat disturbing to others. A computer can appear to think and communicate like a human, can be a link to other humans, can feel like an extension of the self, and can be the basis for strong emotional attachments. To investigate this idea, we asked 100 people if they attributed human characteristics to their computer. Many respondents said they talked to their computer, sometimes as if the computer were another person or as if they were talking to themselves. A dozen people used a name or attributed a gender to their computer. Four people said they referred to the computer as alive, but assigned it no gender. Several of the names they used seemed linked with the brand of computer (e.g., Puter, Mr. Computer).

One avid computer user compared computer software to a person's mind. He said the computer was all logic, but one can alter the logic, so it "sort of has a personality." He sometimes preferred computers to people because they were "eight times easier to work with." He often used the computer with a friend he met at a computer store, so they spent many hours working together. Even when alone, "it's like two people are there really." He explained, "I don't just use it, I like it a lot." One mother observed that her children talked about "computer stuff" quite often and even talked to the computer. "You see people talking to it." The children try to figure it out, and "might attribute a thinking ability" to the computer. "They want to know the reason why it does things . . . It is different from people because you can figure it out."

Because children are particularly likely to treat or think of computers as alive, the prospect of making computers look alive may increase confusion. In the mid-1980s, for example, Apple Computers released a mouse to be used with their computers, along with a mouse house, and mouse cover complete with eyes, ears, nose, whiskers, and a colored bow on the top of its head (a pink bow for girls and a blue bow for boys). While holding the pet, the child could press its soft rump to make the hard, cold computer respond.

Today, the computer has become more an extension of the self, while the Internet has become the other. People can develop an active relationship with the Internet where there is a fine line—a portion of a fiber optic, for example—between intelligent life. Some computer specialists have recognized the desire some people have for interacting with machines as though they are interacting with other humans. There are programs that go back twenty years, such as ELIZA, which asks questions that expect reflective, self-analytical responses. Today, the same kind of programs operate on web sites. You can find advice for any problem, have your fortune told, and let a web site help you probe a problem. By interacting with a program that seems individualized and acts like a human, people are able to communicate in human-like ways. Although the trend may be unsettling to some people, it is obviously attractive to others. In the film *2010,* when the computer genius who designed the talking and thinking computer HAL leaves him to "die," tears stream down his cheeks. In the film *AI,* we feel empathy toward the little boy—a thinking and feeling machine—who verges on being human. Our computers and the Internet can seem like, represent, and connect the self to the self or a friend to a friend. Although we don't yet have artificial intelligence, computers have, to some extent, the ability to think. If time comes when humans perceive computers as living beings, interpersonal communication will be redefined.

The affiliation between the field of psychology, sociology, and communication indicates a relationship between the concerns of how people interact with one another. Thus, it is not surprising that researchers in these fields provide the most concrete research shedding light on understanding the computer's effects on human interaction. One must understand the computer's effect on individuals to understand how it influences their communication behaviors. Do computers and the Internet free people of work or free work of meaning? Do you feel like work becomes play or work becomes some meaningless, endless, inputting of data?

Interactive Communication

The Internet creates cognitive and social interaction. The interplay of that cognitive and social interaction can intensify mental or cognitive collaboration.

Mental effort or concerted thinking is difficult, time consuming, and typically avoided by most people (Petty & Cacioppo, 1986). We tend to avoid difficult thinking, not because we are lazy but because of our cognitive limits. Internet-mediated communication expands individual boundaries. The Internet offers pathways that help serve as guiding principles to acquiring information, maintaining relationships, and more. The extensiveness of the Internet exerts a powerful influence on the user because there is always more to learn, know, and discover.

The Internet is a human process of social interaction which affects language and symbols. Unlike other media, the Internet requires the user to interact with oneself and others. A person can sit back and watch the news or listen to music from the Internet, users also read, write, play games, talk to others, and engage in an interplay of ideas and fantasies, so it is no coincidence that people attach the word "interactive" to many computer programs. There are parallels between Internet-mediated communication and the concepts of interaction involvement (e.g., Cegala, Savage, Brunner, & Conrad, 1982). *Interactionism* is a theory of communication that proposes that the process of social interaction affects and changes language and symbols. Communication acts as the essence of society and the major force of social life (e.g., Andersen & Guerro, 1998; Littlejohn, 1977). Cegala (1984) suggested that people with high interaction involvement could better process information, recall details, and had stronger egos. Can the same be said of highly involved Internet users? Cegala said that low involvement people feel uncertainty about interaction with strangers. Can we assume that people who use the Internet to meet strangers are high involvement people? Or does the Internet intensify high interaction involvement?

Through cognitive collaboration, Internet interplay intensifies an interactionist effect on self-development. Some people at least temporarily develop into very different people on the Internet or perhaps a better way to say it is that some elements of an individual become intensified while communicating on the Internet. Part of the explanation for this intensification is that the Internet can increase a person's imaginative interaction with the self. The interactionist theory appears valid in interpreting interaction on the Internet.

Role-Playing Identity

On the Internet, people engage in significantly higher levels of spontaneous self-disclosure than they do in face-to-face discussion. This self-disclosure is probably because in CMC, people are more aware of themselves and less aware of the people to whom they are talking (Joinson, 2001).

> In communication, which is the primary activity, knowing the identity of those with whom you communicate is essential for understanding and evaluating an interaction. Yet in the disembodied world of the virtual community, identity is also ambiguous. . . . In the physical world there is an inherent unity to the self, for the body provides a compelling and convenient definition of identity. The norm is: one body, one identity. (Donath, 1999)

On the Internet, however, that norm is violated because you may not know with whom you are speaking. Some people may become friends and meet face-to-face. Some lists, for example, have gatherings where members can visit each other. Some people exchange photos. Often, there is relatively little information about a person, and that knowledge level affects communication. To most people on lists, however, an anonymous e-mail lacks credibility. In other cases, the Internet provides an opportunity to pretend to be someone else and to experiment with identity. Some people seem to experiment with their identities and manifest Dr. Jeckyll and Mr. Hyde characters online. This imagining of the self can create paradoxes for the self (McRae & Cree, 1999). The relationship between the self, identity, and online interaction is intertwined.

Some interesting questions arise:

- Is the nature of being human based so fundamentally on interpersonal relatedness that the Internet poses basic threats to our being?

- Or, does this technology expand the human abilities and facilitate the emergence of alternative identities that might otherwise remain trapped, encumbered, or dissociated?
- How, for example, does the Internet serve to blur femaleness and maleness into a more neutered, techno-gendered e-mailness?
- And why have so many people come to favor these cyber relationships over face-to-face interactions? (Civin, 2000)

For the person with disabilities, the Internet's disembodiment creates freedom. A woman who is a quadriplegic, for example, told us that the Internet allows her to have normal conversations and friendships with people all over the world. If a person saw the woman first, they may be put off by her paralysis. Through the Internet, she can create a relationship before anyone passes judgment about her because of her disability. Thus, Internet-mediated communication actually has "advantages over face-to-face communication for people with disabilities" (Fox, 2000).

Culture

Some scholars suggest that learning communication is the same as learning culture, in that communication styles and expectations are specific to each specific culture. *Culture* is the mores, social principles, and values of a group of people. Culture may be cemented by geography, ethnicity, nationality, communication patterns, or other factors. Let's examine three United States cultural values that correlate with how people use the Internet.

People of the United States value their cultural diversity. We used to call the United States the *melting pot,* suggesting that we were a land where many people came together to become one. Now we use the metaphor of the *stew* or *tossed salad,* still many people blended together but keeping distinct flavors of their own. The World Trade Center represented the enterprises and economic cooperation of nations from around the world with the many peoples of the United States. When the World Trade Center collapsed, people from dozens of different nations died. The U.S. value of cultural diversity is reinforced by the World Wide Web, which is an interlinking of people around the world. With much of the Web being English-language and U.S.-hardware and software based, the Internet represents a kind of cognitive U.S. American diversity.

Citizens of the United States value their individuality. Internet exploration is an individualistic activity. A person sits at a computer or holds her or his smart phone and engages as an individual. Although the person may be connected to other people, that is through the medium of the Internet, which then allows a kind of mental collaboration. The individual activity enables the person to pursue people, contacts, and information in a highly personal and individualistic way.

U.S. citizens value the work ethic. If you consider some of the often repeated sayings—"Never put off until tomorrow what you can do today"—you will see embedded in this culture an importance of work. Many Americans derive their identity from their job. And, U.S. workers tend to have less time off than the work force of other comparable nations. In the United States, there's a sense of "get over it and get on with it," typified by the saying "Pull yourself up by the bootstraps," which means take care of yourself, solve your problems, and forget the past.

First, that work ethic causes some people to feel guilty or frivolous when they are at leisure. In fact, U.S. workers have less vacation time than is typical for other comparable nations. As part of that work ethnic, the ability to collaborate cognitively is enhanced through the Internet. And there is a blending of play and work. So, often people stay connected to work on the weekends, holidays, and vacations: "You can contact me via e-mail while I'm on vacation."

Age

You may have heard many warnings about the negative influences of computers and the Internet on children. Predators use the Internet to lure young victims. Children can access pornography and hate speech unless their parents install software that limits access to safe sites only. Yet, despite the many dangers lurking in the shadows, the Internet allows children to access information and talk to people online. They can play and collaborate in ways they cannot do in other contexts. Computers affect the way children learn and process information. The way they play when a computer is involved is different from other types of play activities because of the specific task demands required by computer use. The use of the computer can, in fact, foster representational ability and thinking in children (Fletcher-Flinn & Suddendorf, 1996b). Does the Internet warp young minds? The research suggests the Internet is a relatively safe way to investigate the world and move into adulthood.

> As the director of MIT's new Comparative Media Studies Program, I had been called months earlier to testify before the U.S. Senate Commerce Committee hearings on "marketing violence to youth." As the father of a 19-year-old son, I already knew how contemporary adolescents were using digital media to expand their social networks and how important those links could be, especially for outcasts or kids at risk. Trying to better understand youth perspectives, I launched a tour of American high schools and monitored teen web sites. Those experiences convinced me that many of our kids are going to be all right, not in spite of the fact that they are growing up online but because of it.
>
> Teens themselves often describe the Web as a utopian space, a refuge from divorced parents, economic hardship, crowded classrooms, intolerant teachers and hostile peers.
>
> Many kids feel they have little say in their real-world environments, and so they value the Web as their own world—where they set the rules. (Jenkins, 2001, para. 3, 4, 5)

When youth use the Internet to play, to imagine a personal world, or to reach the real world, they engage in an interplay with others on the cognitive and emotional level. The Internet is a relatively safe way to investigate the world and move into retirement as well. Many seniors appreciate the mental stimulation and cognitive connection they can accomplish through the Internet. Seniors may take a little more time in learning equipment, but in general, they are embracing technology. The beliefs that older people of the United States lack the inclination to use the computer are clearly prejudicial in light of actual adoption and ownership statistics. For people living in assisted living or nursing homes, the Internet provides a way to go beyond the isolating boundaries of the facility, especially in connecting to people of different ages. Like everyone else, seniors use the Internet primarily to communicate with family and friends. Intergenerational learning has been touted as a success (White, McConnell, Clipp, & Bynum, 1999), a finding with which we agree. One author of this book has met with success in teaching an intergenerational Internet course where participants range from age 17 to over 80. It has been suggested that a primary reason for lower computer use by older adults (approximately 55 years-old and up) is simply because of the lack of appropriate instructional materials (Morrell & Echt, 1997).

Though online older adults are proportionately underrepresented as a group, they are steadily increasing in number. The eMarketer[10] research shows that the population of seniors online has grown by over a third between 2000 and 2001, from 6.6 million to 9.1 million. This parallels an overall increase of people over 55 in the general population, from 60 million in 2000 to 66 million by 2005—according to U.S. Census Bureau figures. The 15 percent of U.S. seniors who are online are enthusiastic net users. They spend an average 8.3 hours per week online, which is more than any other demographic group. College students use the Internet for 7.8 hours per week, adults for 7.7 hours, and teens only spend 5.9 hours per week online. At the same time, older adults are lagging in Internet usage. There appears to be a relationship between age and being in the labor force. Looking at age alone, individuals who are over 50 years old are among the least likely to be Internet users (29.6% in 2000). In August 2000, however, the rate for individuals age 50 and older who were still in the labor force (46.4%) was much closer to the 58.4 percent for 25 to 49 year olds who were in the labor force ("Falling through the Net," 2000). But we are particularly interested in older adults who are out of the labor force and their use of the Internet. The reason for our interest in this group is that it seems reasonable to hypothesize that this particular group has much to gain from using the Internet.

We think it makes sense that seniors will benefit greatly from logging on. Web sites such as SeniorNet provide a multitude of communication opportunities that include roundtables on a variety of recreational and information topics, places for socializing, and support groups (Nussbaum, Pecchioni, Robinson, Thompson, 2000). Other sites have information on travel, health and hobbies. The possibilities are endless; it is just a matter of getting senior citizens introduced to the Internet and computers. Hacker and Steiner (2001) reported that "Internet usage frequency is significantly related to the usage hurdles of skills, opportunities and comfort and that Internet usage frequency is significantly related to the benefits of using the Internet in the areas of jobs, financial information, interpersonal communication, and political information" (p. 399). Interestingly, seniors make up a large part of the population. Since 1900, the percentage of people in the United States over the age of 65 has tripled and the Administration on Aging (1998) has reported that people 65 years old or older represented nearly 12.7 percent of the U.S. population. By the year 2030, 19 to 21 percent of the total will be over age 64.

Traditionally, older adults have not been large consumers of educational services, however, technology changes rapidly and it will be crucial for older adults to adapt in order to cope successfully with the changing world around them. That is, many seniors will want to use computers simply to stay socially integrated, and this is a very good thing. Nussbaum, Pecchioni, Robinson, and Thompson (2000) sum up their book, *Communication and Aging,* with these words: "As a final statement, we believe that the most important point of the book is that beyond the physical and financial, successful aging is essentially a relational entity, dependent on mutually satisfying and functional relationships, maintained and defined throughout an individual's life span by competent communication" (p. 342).

According to Hacker & Steiner (2001), Internet use is not simply a function of access to the Internet in terms of having a computer or access to a computer (also see van Dijk, 2000). Instead, they report on data that suggest that Internet use frequency is also a function of computer skills in using the Internet, opportunities to participate in online communication, and comfort with the technology.

The above suggests that more older adults will benefit from using the Internet. There are already signs of this happening, of workshops and courses being offered specifically for this purpose, and of attention to this growing market.[11]

Gender

Whom do you think uses e-mail more, women or men? Whom do you think has less stereotypical attitudes about the Internet, men or women? Generalizations are always problematic, including those based on gender. We all can think of exceptions to every generalization, but one study found that the answer to both questions was women. Women use e-mail more and have less stereotypical attitudes about the Internet (Jackson, Ervin, Gardner, & Schmit 2001). Traditionally, research suggested that women have limited influence on technology (Teich, 1997; Wajcman, 1991). The research contended that women never have been appropriately credited with inventions because they needed financial supporters who were men. It contended that a woman's approach to technology tended to be subjective, and she lacked knowledge and experience of old ideas required in technology.

Today, a woman can communicate effectively on the web, including designing web sites with little money or expertise in technology. In light of the speed that knowledge changes, old ideas can fail to maintain meaning in technological development. Thus, research about women using technology just a few years ago lacks validity with current user-friendly computers, free services, and intuitive programs. Today, a woman as well as a man can use a local library to access the Internet, where she can access her free e-mail account, moderate her free e-mail list, and maintain her free web page accounts.

One of the many forms of help is that women direct each other to some of the thousands of women's web sites and hundreds of mailing lists on the Internet. If a woman mentions a particular interest while talking on one list, other women will suggest lists she might try. If a particular list requires a referral by a woman who is already a member, a woman may enable someone she likes to enter another list.

The result of these developments are enabling a wide spectrum of women to use online communication, an activity they find rewarding, enjoyable, important, and empowering. We have heard women talk of the Internet as their connection to other women, which enables them to generate strong emotional attachments.

One study found that males and females took distinctively different roles in the online interaction. Males sent (on average) more messages than females, a finding that contradicts more recent research. Men wrote messages which were twice as long as those sent by females, and made more socio-emotional contributions than females. Females, however, were found to contribute more interactive messages with group members than males (Barrett & Lally, 1999).

In another study of online interaction, women were more private (disclosed less) and less talkative than men (Adrianson, 2001). Whether or not a person can correctly identify the gender of the other person in an online encounter is a complicated matter that depends on many variables. In general, assume that you cannot be sure whether the other person is a female or male (Savicki, Kellye, & Oesterreich, 1999).

Men tend to expect that their use of the technology will have the outcome they desire more often than women do (Comber, Colley, Hargreaves, & Dorn, 1997; Fletcher-Flinn, & Suddendorf, 1996a).

Reflection

Does Internet interplay intensify your intrapersonal processing? Or does it dull your senses? What are the paradoxes of cognitive collaboration on the Internet (see Table 7.1)?

Does the Internet happen within your self? Yes, but to some extent that communication happens because of social interaction and the attachment of meaning. The generative ideas of symbolism and meaning apply broadly to every aspect of human communication: to education, interpersonal relationships, persuasion, motivation, development, interpretation, and expression. In this view, the Internet extends, intensifies, and makes visible the central feature of relating individuals to symbolic stimuli.

Cognitive collaboration occurs when two people share ideas so that they become of similar minds. Have you experienced cognitive collaboration? Can you give an example of how Internet interplay can intensify immersion into another person by creating a sense of intellectual communication or mental connection.

TABLE 7.1 *Paradoxes of Intrapersonal Communication on the Internet*

Intrapersonal.	Mass and generalizable.
Inside the self.	With the many.
Subjective.	A system that operates with certain rules and laws.
Individual interpretation of the Internet.	Mass effects of the changes in communication caused by the Internet.
Enable people to celebrate their differences.	Make the world the same and individuals part of a global village.
Enhance individuality.	Loss of individual cultures.
Emphasis on text content.	Emphasis on visual medium.
Enhances resources for person working alone.	No one works alone any more because the Internet imposes a system of cognitive collaboration.

Cognitive collaboration is the interplay of the inner and the outer, your intrapersonal and interpersonal communication. Cognitive or mental aspects of Internet interact so that people reach a common meaning with others on an intellectual level. What is your cognitive collaboration through the Internet?

Case for Discussion

In your course discussion group, give your analysis and perceptions on the following.

Among children, there seems to be no gender difference in liking computers (Nelson & Cooper, 1997). Although some research suggests differences between female and male Internet communication because of fear of technology, other researchers suggested that the entire nature of computer use and involvement has changed the nature of computer anxiety (Thatcher & Perrewe, 2002). Their research demands that we question previous research correlates with age and gender variables that previously suggested that older people and women were more anxious and less successful in computer use than younger people and men. Engage in a discussion about this topic.

Internet Investigation

1. *Join an E-Group.* If you are a man, sign up to participate in a discussion group for women. If you are a woman, sign up to participate in a discussion group for men. Observe the interaction. Write anecdotes that illustrate the nature of the Internet discussion, which might be different by members of the other gender.
2. *Mass Communication View of Online Journals.* As you consider the ideas of this chapter, how might you see the ideas through the eyes of an expert in mass communication? In what ways does intrapersonal communication become mass communication through the Internet? Weblogs—a type of personal sharing on the Internet—are a combination of intrapersonal and mass communication. How might an individual's intrapersonal processing change when verbalized publicly? If intrapersonal information, ideas, or processing are provided to others are they still intrapersonal? When the Web is involved, does intrapersonal communication automatically become mass communication? Does anything about our understanding of the communication process change when we consider the communication from the standpoint of intrapersonal communication versus mass communication?
3. *Online Discussion with an Orangutan?* No, but for years, the National Zoo in the District of Columbia has operated an orangutan think tank. One aspect of the think tank teaches orangutans language acquisition through touch-screen computers. Although the orangutans are not ready for instant messaging with humans, they do show remarkable capacity for using the computer. Investigate ways in which the human-computer connection might includes animals.
4. *Self Paradox.* Scholars argue both sides of the paradox related to self and the Internet. Some researchers believe the mass media expose us to so many different cultures that we have a difficult time creating a stable sense of self (Gergen, 2000).

Others say the Internet cannot affect the self, although technology opens new opportunities for individuals (Wynn & Katz, 1997). Do you think the use of the Internet affects the self? What does a person say about herself or himself by using a signature file? In your investigation, consider possible influences, such as participation in MUDs, role-playing, imaginary personalities, character play, fantasy sex behaviors and how they might affect the self.

5. *What's in a Name?* As you analyze the Internet, consider the role of names in meaning-making. If we assume that names are important to the sense of self, how do screen names, pseudonyms, aliases, and nicknames affect the self? Why is the use of personal profiles important? How do nicknames affect the nature of a person's communication online? Are they more open, for example, because they believe they are anonymous? Find a web site that provides information in finding the real name behind a particular nickname, such as http://www.sobs.soton.ac.uk/email/eudora.shtml or http://flex.ee.uec.ac.jp/texi/sc/sc_5.html Investigate the topic and share your findings with other members of the course.

6. *Stream of Consciousness Writing.* People use various writing styles when communicating via the Internet. Some never capitalize anything. Others connect all phrases with ellipses (. . .) or hyphens (— — —). The stream-of-consciousness style suggests internal processes being externalized, and provides an excellent example of cognitive collaboration. You may want to work with other students in the class to investigate online writing that reflects cognitive collaboration. Examine group discussion archives, for example, as a way of identifying examples and trends. Share the information with class members.

Concepts for Analysis

We have provided concepts relevant to this chapter so that you can check your understanding. These concepts are worth dissecting in more detail. As a research option, examine one concept. Use our citations as a starting point for your study. We recommend that you search for refereed articles in an online database. Find at least three solid references you can read. Summarize, scrutinize, and present your analysis to other people in your course.

Questions to stimulate your inquiry: Do you agree with the ideas as we presented them? What other points of view should be considered? What details need to be added to fully understand the concept? What can you contribute through your personal analysis of the concept? What more do scholars need to know about the topic?

Ambiguity is the lack of predictability in communication.

Cognitive collaboration occurs when two (or more) people jointly explore ideas, communicate encoding and decoding options, discuss intrapersonal processes so that they become of similar minds. The Internet interplay intensifies that immersion into others by creating a sense of intellectual communication or cognitive collaboration.

Communication is reaching a common understanding through interplay.

Context is the situation, environment, or situated meaning. The Internet alters and reduces the relevance of the physical context. In a sense, the Internet transcends the time and place of communication.

Cultural mediational artifacts are the nonverbal communication and cultural information that help people interpret communication. The physical and cultural differences are easily hidden in communication on the Internet.

Culture comprises the mores, social principles, and values of a group of people. Culture may be cemented by geography, ethnicity, nationality, communication patterns, or other factors.

Groupthink is a problem that develops when a cohesive group makes bad decisions. In an effort for collaboration in their work together, the group may begin to take an unrealistic view of the problem.

Your *intention* is your aim or purpose.

Interactionism is a theory of communication that proposes that the process of social interaction affects and changes language and symbols. Communication acts as the essence of society and the major force of social life.

Intrapersonal communication is communication within one's self, decoding and encoding, meaning-making, whether what is decoded originates inside or outside the body of the intrapersonal communicator, and whether what is encoded is actually expressed, leaked, or given off (Figure 7.1).

Juxtaposition of utterances refers to the position of utterances next to one another.

Mediation is consideration or negotiation which takes place on the Internet. Similarities between people can be created in the minds of each individual because Internet users seek commonality, which they can find in their intellect.

Metamessage is the larger, holistic meaning of a message, or the attitude communicated by the message sender or the speaker's attitude toward what is being said.

Mind refers to the cognitive processes of thinking. The Internet intensifies the role of the mind in attaching meaning.

Second self is the idea that the computer acts as a second self by engaging the individual in intrapersonal communication and self-exploration available in no other way (see Turkle, 1984).

Tone is your mode of expression.

Turn-taking is the communication behavior of alternating messages from one person to the other.

Voice is your communication point of view.

Notes

10. http://www.emarketer.com/bin/AT-emarketersearch.cgi
11. *http://www.usm.maine.edu/com/seniorcol/sencoll.htm*

8

Interpersonal Communication on the Internet

During the war in Afghanistan, the USS Carl Vinson, an aircraft carrier, processed 30,000–50,000 personal e-mails per day (CBS television report, December 7, 2001). Every sailor on that ship has access to a computer which she or he can use to connect with family and friends. While away at sea, e-mail serves as a relational lifeline for sailors. This example is just one of many that show how the Internet can enhance communication in an interpersonal context. The context is the place or situation or framework in which communication takes place. Although you could say the context is the Internet, we consider the context to be the framework from which the person communicates. Admittedly, "context" is a fuzzy concept that covers a great deal of ground. Context can refer to the immediate physical surroundings, personal history between people, cultural understandings, gender, or even roles. You might be able to think of others that apply. You can think of each instance as a set of rules or knowledge that can be applied to the communication event in the process of meaning-making. Since context affects meaning-making as people communicate, it has clear implications for interpersonal communication on the Internet.

Focus

```
TO: Students

FROM: paradoxdoc@hotmail.com

RE: Overview: Interpersonal

Lenny and Joan talk about one of the most interesting
and rewarding aspects of communication on the Internet:
How does the Internet affect interpersonal relationships?
Internet communication is more like interpersonal
```

> communication than other forms of communication. Not only is the Internet often a type of interpersonal communication, but the Internet can affect face-to-face interpersonal communication.
>
> You probably already use the Internet to enhance your established relationships, by keeping in touch with friends and family. You may participate in chat, instant messaging, MUDs, or online discussion groups. Not only might you create an interpersonal relationship, but you can also feel a part of a community through online communication. Lenny and Joan talk about speed, anonymity, interactivity, and regard as variables that affect communication online.
>
> You can use the Internet to meet people and create relationships. You can use the Internet to enhance relationships with family and friends. You can use the Internet to maintain long-distance relationships. The Internet—like other earlier forms of communication media—was not originally conceived for interpersonal communication. Yet, in the hands of users, that is exactly what has developed so far. Hence, a key question Lenny and Joan ask you to ponder: Is the Internet good or bad for people's interpersonal lives?
>
> Lenny and Joan ask you to consider the ways the Internet can influence a family—both positively or negatively. The Internet can provide a source of communication content, for example. The Internet can increase human interplay. The Internet can reinforce a child's self-esteem. The Internet enables new opportunities for interpersonal relationships. And finally, Lenny and Joan ask you to explore the nature of online relationships.

Sometimes an individual's personal life and professional life are clearly separated, while at other times the line between personal and business context is blurred. In this chapter, you will examine interpersonal communication on the Internet in the personal context. At the end of this chapter, you should be able to answer the following questions.

1. What is the relationship between interpersonal communication and the Internet?
2. How does the Internet shape family communication?
3. How does the Internet shape the self?

4. In what ways is the Internet a means of creating interpersonal relationships?
5. What are some of the paradoxes of interpersonal communication on the Internet?

Relationship between Interpersonal and Internet Communication

Interpersonal communication is one-to-one communication, and some communication scholars consider the term synonymous with interpersonal relationships. We believe interpersonal communication can be *face-to-face* (f2f or FtF) or between two people online. You may know people who have met online as the prelude to a face-to-face communication. You may have done so yourself. Perhaps you met someone online, thought the person had interesting potential, and decided to meet face-to-face. Maybe you *googled* a blind date, which means that you investigated the person through Google.com or another search engine to find out information before the face-to-face meeting. You have undoubtedly heard about online dating services and maybe you have tried one yourself. All of these examples are ways that the Internet is influencing interpersonal communication and relationships (Howard, Rainie, & Jones, 2001).

You probably already use the Internet to enhance your established relationships, by keeping in touch with friends and family. You may have participated in chat, instant messaging, MUDs, or online discussion groups. These personal uses of Internet communication are the general focus of this chapter. When it comes to relational communication, Internet communication is similar to face-to-face communication (Barnes, 2002). Although a mass medium in many ways, to the user, computer-mediated communication (or Internet-mediated communication) seems more closely aligned with interpersonal communication than with other types of communication (Walther, 1996; Walther, Anderson, & Park, 1994).

So within that interpersonal perspective, this chapter will consider personal contexts, individual purposes, flaming, identity, storytelling, online relationships, and relational stages as a part of human communication on the Internet. Not only is Internet communication similar to face-to-face communication, but face-to-face interpersonal relationships can be influenced by online communication (Wolak, Mitchell, & Finkelhor, 2003).

Interpersonal communication is communication between two people, typically face-to-face. *Community* is the ability of people to come together, to have a sense of sharing and commonality in an online environment, and to feel a sense of gathering or oneness (Ferrigo-Stack, Robinson, Kestnbaum, Neustadtl, Alvarez, 2003; Pack & Page, 2003; Postmes, Spears, & Lea, 2000). *Immediacy* is a feeling of closeness, of sensing an emotional proximity. Although Internet users may be separated by enormous physical distance, they may feel immediacy and community through CMC, which can provide a vehicle for making friends (Parks & Floyd, 1996). Further, our interpersonal communication characteristics probably determine how we communicate online (Palmer, 1995). And the fact that most users create a sense of humanization in their e-mail makes their communication more interpersonal in nature (Swats & Walther, 1995).

Researchers have questioned whether interpersonal communication over the Internet will enhance or attenuate human relationships. Walther (1996) explored the heightened nature of interpersonal interaction online, which he called the "hyperpersonal" nature of computer-mediated communication. Because of the text-based nature of interactions, people may develop relationships online that meet or exceed face-to-face relationships. How is that possible? Because many individuals incorporate a fantasy, stereotypical, idealized nature to the relationship, which they perceive as reality. Is that reality or fantasy?

One line of research concerns the idea that a particular communication medium may substitute time online for other forms of communication. Obviously, the concern here is whether or not time on the Internet will substitute for other social and media involvement, such as reading (Kohut, 2000; Kramarski & Feldman, 2000). At this point, it appears that the idea of substituting Internet communication for interpersonal communication F2F does not hold up. If anything, a number of studies actually suggest that Internet users are either unaffected or more involved in social interaction than nonusers (Robinson, Kestnbaum, Neustadtl, & Alvarez, 2000). Research has shown Internet users to be more involved in reading literature, attending arts events, movies, and playing sports than nonusers (Robinson & Kestnbaum, 1999). Experience teaches us though, that as research on a new medium unfolds, we learn to make finer distinctions. In this case, it may turn out that comparing Internet users to nonusers is too coarse, that we will need to look more closely at such variables as early adopters versus recent users. Time and research will tell. For now we can say that the evidence does not suggest a deleterious effect of the Internet on interpersonal communication.

In their review of the literature on the social implications of the Internet, DiMaggio, Hargittai, Neuman, and Robinson (2001) sum up the research on the Internet's effects on social interaction and civic participation this way: "Internet use tends to intensify already existing inclinations toward sociability or community involvement, rather than creating them ab initio" (p. 314).

Wheeler (2001) presents an ethnographic study of women's use of the Internet in Kuwait, a conservative Islamic environment, focusing on the question of whether or not Internet communication liberates women in a culture that employs social sanctions to constrain voices. In Kuwait, women in particular are highly constrained in their public behavior, with sanctions imposed for mixed gender interaction and for criticizing the status quo. One issue that Wheeler was interested in was whether or not access to the Internet would be the key to liberation. In a society where women are not allowed to mix freely with men outside of the family or to speak freely, will they take to the Internet and, in addition, will they bring back into their local lives lessons learned there. In short, what she found was that access to the Internet in and of itself was not enough to bring about revolutionary change. Instead, she argues that:

> Cyberspace is an extension of the realms of social practice and power relations in which users are embedded. At times voice is liberated from gender restrictions, like from within the cyber-relations enabled by IRC. But voice is historically subjected to constraints based upon publicly enforced notions of right and wrong in public discourse. The advent of new fora for communication do not automatically liberate communicators from cultural vestiges

which make every region particular and which hold society together. In Kuwait, this means that women are not likely to organize and to speak out against their husbands, their brothers, their sons and fathers, their bosses; this would be to publicly embarrass their patriarchs. It is more likely that voices will be lifted in the privacy of an office with the door shut, or in the living room during hours that many husbands are at work. (p. 202)

We can see how communication on the Internet may influence communication, but, in addition, we must take care in inferring from this just how local, interpersonal communication will be affected. It may be that features such as anonymity online and discussion with people at a great distance do influence how we communicate. Wheeler reminds us that local culture cannot be ignored. Once again, context plays a key role in interpreting communication behavior. Stewart, Shields, and Sen (1998), in their study of students on a course listserv, also found evidence for the effects of culture and context on online communication.

Gurak (2001) offers a short list of features that she thinks characterizes communication on the Internet: speed, reach, anonymity, and interactivity. These key features, according to Gurak, influence communication of all kinds on the Internet, but we can apply them to interpersonal communication in particular. Speed refers to the time it takes to send and receive messages—obviously it is very fast. And our expectations of how long it takes have moved along with the technology. We write quickly as we interact with others on the Internet, reducing writing to some hybrid of speech and writing, abbreviating and ignoring spelling and punctuation to a great extent. Reach refers to our ability to be in touch with people at great distances, and again, at great speeds.

Anonymity refers to the behavior of people who create an identity online, claiming to be someone they are not, manipulating gender, age, occupation, health status, and so on. Interactivity refers to the ability of online participants to not only receive messages, but to react to them. Unlike the broadcast audience, the online participant can actively enter into the communication event. It is not difficult to see how these features would affect who we interact with, how we interact with them, when, and what we have to say. Clearly, speed, reach, anonymity, and interactivity hold implications for our interpersonal communication online. We suggest adding at least one more feature to the list, and that is *regard*. Put simply, think of regard as a message that acknowledges you. What we have in mind here is how a message suddenly appears on the screen, a message that indicates that someone took time to think of you and to write something to you personally, not you as a potential anonymous consumer or a member of some list. Perhaps it comes from someone you can imagine writing to you. You go into your e-mail software or the chat room or the message board and there it is waiting for you. A moment earlier it was not there. You show up and someone knows your name, as the television program "Cheers" used to say in its theme song. We are thinking of that feeling of encountering someone familiar or at least someone familiar with us or our deeds or ideas, someone who takes the time to recognize us and pay attention to us. At any time, a message may arrive that is for you. We can think of this as going to our physical world mailbox repeatedly throughout the day (or night) and finding mail there on an unpredictable schedule. The element of surprise plays here, as well as the element of hope and excitement, that something good or interesting might be waiting for me. Someone out there will send me a message that confirms my existence and possibly confirms something about me. So, for instance, if one is involved in one or more online interpersonal relationships, there is always the potential of a pleasant surprise, of hearing from a friend or romantic partner. The concepts associated with regard that come to mind here are as follows: respect, affection, esteem, looking at, attention, admiration, and to take into consideration. Writing about the individual within the collective in cyberspace, Fernback (1997) says: "The exciting sense of possibility permeates cyberspace" (p. 38).

Meeting People and Creating Relationships

For example, Chris and Pat met in a chat room for people living in their geographical area. They started talking online and decided to meet. They liked each other and began dating, and eventually became romantically involved.

A school teacher told us that her son became friends with a boy six years older, which she attributed to their common computer interests. "Nobody that much older ever talked to him before," so the relationship has given her son a sense of importance. "I've seen a sense of power in some kids . . . kids without any place, who really latched onto the computer."

Meeting strangers online to form casual or romantic relationships has been well documented (McCown, Fischer, Page, & Homant, 2001; Lea & Spears, 1995; Parks & Floyd, 1996; Parks & Roberts, 1998; Turkle, 1995).

We see anonymity, regard, interactivity, reach, and speed at work here.

Enhancing Relationships with Family and Friends

James is 90 years old, a retiree who lives in Florida. His children live in Germany and Missouri. His grandchildren live in Maryland, Virginia, and Michigan. James doesn't see his family members as often as he'd like, but he receives e-mail and updates about the family and digital photos regularly from everyone. In addition, James can read about and see photos from his grandson's high school varsity games through the school's web site. The game summary and photos are online by 8 A.M. the morning following each game, which helps James feel a part of his grandson's life. James and his grandson often converse back and forth on instant messaging about the most recent game. We see speed, regard, interactivity, and reach at work here.

Maintaining Long-Distance Relationships

Jessica and Roy are colleagues who work together online. They communicate daily about work and their personal lives. Over the years they have become friends as well as co-workers. Although they live on opposite sides of the country and have only seen each other face-to-face a few times, they have a strong and valuable co-worker/friend relationship. What makes for effective e-mail that enhances relationships? *E-mail* is electronic mail or messages sent and received via the Internet. They are quickly written and can be read whenever a person wants. Most people expect a quick reply to an e-mail because the ease, accessibility across distance, and speed of e-mail are its primary attractions. E-mails can cause problems, however, because online communication can threaten good relationships because of the lack of control of information and the potential for trust violation. A personal e-mail forwarded to someone without their knowledge or expectation, for example, violates privacy and trust expectations in relationships (Schonesheck, 1997). One woman, for example, told us about writing an e-mail in which she vented her feelings about work to a colleague, then without her consent, the colleague forwarded her e-mail to her boss. She was embarrassed and compromised by the colleague's forwarding of the e-mail.

Content analysis of e-mail messages shows that e-mails are more similar in style and content to face-to-face interactions than to any other type of communication (Newhagen, Cordes, & Levy, 1995). Not only is computer-mediated communication similar to face-to-face interpersonal communication, but the use of the Internet also may affect the way people communicate interpersonally face-to-face (O'Neil, 1996). But we take the analysis of communication on the Internet a step farther after having noticed several recurring factors in Internet communication. First, there is often a *casual* sense of play involved in the interaction. Second, there are often *paradoxes,* oppositions—pros and cons—contradictory meanings and behaviors associated with Internet communication. Third, certain aspects of communication are strengthened or *intensified by Internet* use. Fourth, Internet-mediated communication strongly *engages intrapersonal* communication, and can be seen within the cognitive framework. Fifth, the *attachment of meaning* to the message and the process of Internet communication gives the communication its significance on an individual cognitive and more general social basis.

Generally, the Internet is correlated with positive associations. Although there are paradoxes that suggest negatives—some with violent and deadly consequences—we observe a great deal of healthy Internet interaction in the personal context. And we agree with researchers who find that typical Internet users are generally healthy and effective communicators. For example, the Internet user seems to be "smarter, healthier (that depression study was debunked), and more physically active" than are Internet-nonusers (Golson, 2001, p. 14).

Individual use of the Internet is as varied as there are people using the Internet. Children, the elderly, and everyone in between communicate on the Internet. On a personal level, people use the Internet to connect with family and friends, to provide entertainment, and to access information. Individuals use the Internet for social connection through discussion groups, support groups, health information, to promote rehabilitation, explore hobbies, study for lifelong learning, and as a way of receiving services. People of all ages use the Internet in these many ways. Consider this array of findings (Shedletsky, Ouillette, & Monfort, 2001) from two groups of senior citizens who were taking a course introducing them to the Internet. The groups, with an average age of 69 and 72 respectively, gave the following reasons for wanting to learn how to use the Internet when asked on the first day of their Senior College class, "What was your reason(s) for taking this course, Internet for Seniors?":

Group I
1. To gain skills in using the Internet.
2. To increase knowledge of e-mail (also computers in general).
3. To break through—(I was an editor/researcher—knew the old way—still find the Internet not detailed enough as say the library for research).
4. To learn more about the Internet.
5. Buying a computer and learning how to use it.
6. To learn as much as I can to expand my horizons.
7. To be able to use and understand computers.
8. To be able to do some things my grandchildren can do.
9. To be at home with my own computer and do research.
10. To learn what I should select for best results.
11. To increase my knowledge of the computer—this is my first course.
12. To understand what is going on in this world of computers.
13. To be more effective using the computer.
14. To learn about accessibility and best routes to take.
15. To know more about the actual use of the Internet—need more "how-to's."

Group II
1. I have great pleasure and learning from the Internet.
2. So that I can be "with it!!"
3. I thought that I might be able to pick up some more Internet experience and maybe get some pointers.
4. Expand knowledge of capabilities of the Internet.
5. I get frustrated when getting information off the Net. I seem to go around and around.
6. To learn basics—Internet and e-mail.
7. To learn as much as my grandchildren know.
8. I feel I should know more about it.
9. To become a little more comfortable with the PC.
10. Frustration with my ability to find what I was looking for.

Interestingly, looking at the reasons these senior citizens gave on day one of their course for learning how to use the Internet, you see a lot of focus on the computer. Like so many of us, these people soon found that the computer could be viewed less as a tool in its own right and more as an instrument of communication. One thing that does stand out when we observe how people actually use the Internet is that a key use of the Internet is for interpersonal communication. Having worked with the senior citizens in the Internet for Seniors course, we can say that the ability to send and receive e-mail, especially to family members, turned out to be a key motivator for this group of newcomers to the Internet. The Internet, like other communication media before it, was not originally conceived for interpersonal communication. Yet, in the hands of users, that is exactly what has developed so far. Hence, a key question for us to consider is whether the Internet does good or bad for people's interpersonal lives.

Skill Section: Speed, Reach, Anonymity, Regard, and Interactivity

We expect the speed of our communication technologies ("Nightline: Brave New World," 1999). Once you have experienced a faster machine, be it computer, telephone, Palm Pilot, or fax, you probably have difficulty when asked to go back to the slower one. We expect *speed.* Have you had that experience of using a faster machine and then using a slower one and finding that you are impatient? Have you ever used a slower computer than the one you are used to and you find yourself tapping your fingers waiting for a web page to load up, or you wonder if the computer is working because it takes "so long" to load the page? In fact, it may only be a matter of seconds' difference, but we become impatient.

As the narrator in Nightline's discussion of our tools asks: "Why, if we have faster and faster and faster machines do we feel like we are running out of time?" Research suggests that people will wait an average of about ten seconds for a web page to appear on their monitor before they give up and move on (Flanders & Willis, 1996, p. 70). People cringe at the thought of waiting through the dialing process of a rotary phone once they have used a touch-tone phone. Do you recall a time when you had to get up and go to the television to change the station, or are you a child of the flipper era, with remote controls that allow you to skip back and forth between stations or that allow you to view a picture in a picture? Time-saving devices, eh? And yet, when we ask our students if they find that all this speed has given them additional leisure time, they strongly react with just the opposite, namely, they feel like they are running on a tread mill and cannot keep up with all that is coming at them. Our students report feeling nervous and rushed, inadequate in keeping up—overwhelmed. While speed may account for some of the thrill and intensity of being on the Internet, the speed may be adding to our harried lives and even to the occasional fits of anger that explode online and off, a kind of Internet superhighway road rage.

Along with the speed of the new technology, consider the *reach* of the Internet. While reach is often thought of in terms of your reaching people who are far away, there is another sense of reach that applies equally here. Reach also refers to the idea that you can be reached just about anywhere, anytime. When a gentleman brought his cell phone into a sauna, just in case the restaurant he managed needed to contact him, this idea of reach was vividly portrayed. "Being off" is something we are losing. Anyone who has taken an online course or even has experienced an online discussion supplement to a course, has found that the convenience is offset by the never ending messages that can reach them. Teachers and students both complain of the added work when they are connected to a course through the Internet.

Jones (1997a) suggests that in spite of the prevailing metaphor of space to capture the Internet experience, it is the Internet's *bias toward time* that is its principal feature. Jones points to our obsession with efficiency and how this concept is shaping our social lives:

> The Internet's bias toward time, on the other hand, marks it as the latest in a series of mechanical developments arising from "the demands of industry on time" (Innis, 1951, p. 74). It is part of a process that has intruded into our everyday life, into the social (see Lewis Mumford's writing for poignant examples), that demands efficiency and results in frag-

mentation and what Innis termed an "obsession with present-mindedness" (p. 87) and what Jeremy Rifkin (1987) calls "the new nanosecond culture." Perhaps its best description, and one that links the Internet's bias toward time to computing generally, is that of a software engineer who stated, "real time [is] no longer compelling" (Ullman, 1997, p. 133). (Jones, 1997, p. 12)

What comes to mind here is the feeling that people are in the way as we wheel our carts through the supermarket, where there is no time to leisurely talk with the workers. We have trouble standing in line for more than a few moments. That we or the next driver is in the way on the highway, that the speed dialer on our telephone is taking too long, or we just can't get enough done in the day. No wonder the people we encounter are upset that others are not taking the time to really listen to them. Is it possible that the Internet is just one more step in the fragmentation of society?

For many of us, the idea that e-mail might be waiting, hopefully something pleasurable, what we are calling regard, adds a kind of excitement to looking to see what is in our electronic mailbox. Not surprising, humans hunger for regard, in a world that so often treats us as a cog in the efficiency model. But it also adds to that sense of speediness, the sense of not being able to process information fast enough, or not having the time to take the other into regard. Some people manage the large amount of e-mails they receive from groups they belong to, often deleting unread messages based on subject lines or author. Some researchers have speculated that this sense of constant connectedness, the speedy way in which we receive and send messages, and the difficulty of keeping up with the information that is reaching us through new and converging technologies, cell phones, Palm Pilots, facsimile machines, computers, and beepers, are causing us to feel nervous and frustrated—to feel rage at times (Gurak, 2001).

Accordingly, it is thought that this increased tension and frustration accounts for road rage both on the streets and highways and on the Internet. Some attribute violence between people in everyday encounters face-to-face as well as angry e-mail between individuals online to the frustration associated with the speed and reach of the Internet (Gurak, 2001).

The term *flaming* has become well known and refers to harsh language directed at an individual online in an e-mail, instant message, or message on a bulletin board. There is some controversy over just how much flaming actually goes on (Dery, 1994; Lea, O'Shea, Fung, & Spears 1992). While we are not sure how much flaming occurs, it does appear to us that flaming may have much to do with just where you go on the Internet. Sticking with scholarly web pages and professional discussion groups, online talk with friends and relatives produces little evidence of flaming, but unfortunately, we cannot say it produces none. Nasty, personal attacks on someone or aimed at something they wrote do occur. Baron (2000) suggests that the individual user's profile is a key factor in predicting flaming (p. 239).

Interactivity online gives us the opportunity to speak back. If one is angry this interactivity is tantamount to the opportunity to express anger. Anonymity online likely enhances the chances that one will express their anger openly, since your anger is not attached to "you" (Gurak, 2001). Internet researchers have suggested that other features of the online medium contribute to flaming. In particular, scholars have speculated about the lack of social cues due to the fact that generally we cannot see one another online (Berry,

1993; Heim, 1993; Kraut, Patterson, Lundmark, Kiesler, Mukopadhyay, & Scherlis, 1998; Slouka, 1995). Hence, nonverbal cues are diminished, reducing the channels of information normally used as feedback in communication. Perhaps the reduced social cues result in a greater tendency to project one's own hopes and fears onto the event, or reduced social cues may render online relationships less fulfilling than f2f relationships (Turkle, 1995). Though, it must be noted that still others speculate that Internet relationships can compensate for reduced social cues online and that differences between online and offline relationships are reduced over time (Lea & Spears, 1995; Parks & Floyd, 1996; Walther, 1992, 1995, 1996).

Some theorists have speculated that the reduced social cues leads to a sense of intimacy quickly and without the emotional investment in f2f encounters (Slouka, 1995). Some writers refer to this view as the social context cues theory (Sproull & Kiesler, 1986). According to social context cues theory, online relationships are not as intimate as f2f relationships. A closely related idea, social presence theory (Rice & Love, 1987), maintains that "social presence" is the feeling that your partner in communication is involved in the communication event. Since social cues, such as physical context, voice and other nonverbals are not available, the communication is impersonal and not as intimate as in f2f exchanges—the feeling of involvement is low. Both the social context cues theory and the social presence theory predict a diminished quality for online relationships compared to f2f relationships. According to social penetration theory of Taylor and Altman (1987), intimacy in f2f relationships, both friendships and romance, is developed slowly as participants reciprocate self-disclosures and probe outer layers of personality before moving to inner layers of biographical information, tastes, beliefs, values, dreams, and concept of self. Relationships that move too quickly, according to social penetration theory, are subject to dissolving. Cooper and Sportorali (1997) have argued that the high speed of online relationships, where self-disclosure exceeds the speed limit for f2f relationships, may account for the sense of intensity. They speak of the relationship being exhilarating at first and eroticized, but also not able to be sustained due to lack of well-established trust and knowledge of one another. Turkle (1995) discusses just such a scenario, where romantic relationships developed in MUDs may display intensity early on, intimacy of a kind, the sense of time itself speeding up, intellectual, emotional, and erotic bonding. But, Turkle points out, these electronic relationships often do not move successfully to the real world, the f2f world (p. 207). As one of her informants noted, in the real life situation he had too much information. In the virtual world of the MUD, he could project onto his romantic partner what he wanted. Interpersonal communication scholars have long since claimed that we construct our relationships (Griffin, 2003). Perhaps interpersonal communication on the Internet intensifies the constructivist process.

Some believe that the reduction in social cues accounts for a tendency to be more disinhibited online, both in self disclosure and in aggression (Lea, O'Shea, Fung, & Spears, 1992; Parks & Floyd, 1996). John Suler (2002) writes about the "disinhibition effect," which covers both flaming and the other side of the coin, particularly generous and humane and open comments made online. He suggests that both sides may be explained by features of online communication such as anonymity, interactivity, invisibility (you cannot see me), lack of immediate feedback, asynchronicity ("running away after saying something"), dissociation or thinking of interactions online as taking place in some imaginary space disassociated from ordinary reality—as if it is all a game. In the imaginary space, one may have the sense of equal status, imagining the other speaking and even "hearing" their voice in our own head. Suler elaborates on this point about introjection:

> Absent f2f cues combined with text communication can have an interesting effect on people. Sometimes they feel that their mind has merged with the mind of the online companion. Reading another person's message might be experienced as a voice within one's head, as if that person magically has been inserted or "introjected" into one's psyche. Of course, we may not know what the other person's voice actually sounds like, so in our head we assign a voice to that companion. In fact, consciously or unconsciously, we may even assign a visual image to what we think that person looks like and how that person behaves. The online companion now becomes a character within our intrapsychic world, a character that is shaped partly by how the person actually presents him or herself via text communication, but also by our expectations, wishes, and needs. Because the person may even remind us of other people we know, we fill in the image of that character with memories of those other acquaintances.

Suler (2002) suggests that online, where status differences are not clearly marked as in f2f encounters, people may be more apt to voice their opinions, not fearing the disapproval of authority figures. He writes: "People are reluctant to say what they really think as they stand before an authority figure. A fear of disapproval and punishment from on high dampens the spirit. But online, in what feels like a peer relationship—with the appearances of "authority" minimized—people are much more willing to speak out or misbehave." In his book, *Virtual Culture: Identity and Communication in CybersSociety,* Steve Jones (1997) asks: "What is it about life offline that makes us so intent on living online?" (p. ix). It may be just those features of online communication that Suler speculates constitute the disinhibition effect which makes life online so attractive to so many people.

Others have speculated on the role of gender (more precisely, biological sex) in flaming, pointing to males and a male style of communicating (Herring, 1993). Gurak (2001) maintains that flaming is more often a male style of interacting online than a female style. She offers examples of males using harsh language and females being supportive or deflecting strong emotions with the use of smileys [☺] or other emoticons (Gurak, pp. 72–74). Stewart, Shields, and Sen (2001) analyzed discussion on a course listserv among males and females, with some diversity in cultural background, and reported both gender and cultural differences. They did find some indication of a dominating style of handling conflict among the white American males.

Influence on Family

A family was discussing who should use the computer one weekend while getting ready to attend religious services together. The parents already had told the 8-year-old daughter that she could not use it because both parents needed to use the computer. "I know," said the wife to the husband, "you go to the early service, and I'll go to the later. That way we can each use the computer while the other is gone." A few minutes later, the wife realized the daughter was hurrying to dress so she could go with her father to the religious service.

"Wait a minute," said the wife, "I was only kidding. When we stop doing things together so that we can have a turn at the computer, something is wrong" (Aitken, 1987). Yet for some families, the computer is no joke, because the lure of the computer can come between family members. More than one woman, for example, has dubbed her partner's computer "the other woman." The computer can have positive and negative effects on families (Kraut, Kresler, Boneva, Cummings, Helgeson, & Crawford, 2002). What is the nature and frequency of family communication associated with computer and Internet use? Families are affected by the Internet and the way they communicate on the Internet (Papert Negroponte, 1996).

Perse and Dunn (1998) found that family members used home computers for entertainment, escape, habit, and to pass time. There has been relatively little study of how computers affect family life (Watt & White, 1999), but the computer appears to affect different family members differently. Some scholars perceive the potential effects on families as negative (Straudenmeier, 1999). We think some steps can be taken to enhance family communication.

Family members can take specific steps to ensure that their Internet use enhances interpersonal communication. Family members can increase effective communication by using the Internet to: (a) provide communication content, (b) increase human interplay, (c) improve individual's self-esteem, and (d) enable new opportunities for interpersonal relationships.

The Internet as a Source of Communication Content

Research demonstrates that computers provide a topic people can discuss (Aitken 1985, 1987). In mass communication theory, a similar idea has been proposed in the name of the *Agenda Setting Theory* of McCombs and Shaw, the idea that the media influence us not so much in what to think but in what to think about, what we attend to (McCombs & Bell, 1996). In a time when interpersonal communication and the family unit may be at risk, families can share information and teach each other about their computer and Internet investigations. Families can make sure they spend time together discussing major computer expenditures. Families can discuss what kinds of auxiliary equipment and programs they desire, and the costs involved. Not only can they discuss computer hardware and software, but the Internet gives an opportunity to talk about logic, ideas, learning, politics, information, and people. Of course, family members can talk about these subjects without being connected to the Internet, but the Internet provides an additional stimulus.

Using the Internet to Increase Human Interplay

Families may want to establish rules to ensure appropriate interaction and Internet use. They might work on the computer together, the parent helping the child with homework. If they are a multiple computer household, they can locate computers next to each other in the living room rather than in separate rooms, so they can converse during parallel computing. Just like with television viewing, the parents may be concerned about excessive violence, obscene language, and explicit sex on the Internet. Parents may want to be in a position to monitor and discuss what the child is doing on the Internet.

One family told us that they have four computers in their household, one for each family member. Three computers are all located next to each other in the living room, and the other computer is a laptop. Although the computer and Internet interest of this family may be atypical, they recognize that their interest can serve as a common bond and point of fun and interest for family interaction.

Families can use the Internet for a vehicle to increase communication (Aitken, 1987). Family members, for example, can leave messages for each other on their computer, work together on a family budget, contribute jointly to the family calendar, search the Internet for a new family car. In one family interviewed, the father wrote programs specifically for his children. The father enjoyed the programming, and the children enjoyed working the programs their father geared specifically to each child's interests. Of course, family members who are geographically distant can enhance their communication through e-mails.

Reinforcing a Child's Self-Esteem

Young children often relate to the computer in creative and fun ways. Children are more apt to interact with the computer as if it were alive. The differences in computer perspectives associated with age can enhance the parent-child relationship. In most cases, the child continually learns from the parent. In the area of computer and Internet-mediated communication, however, the child may be able to teach the parent. One five-year old, for example, learned a graphics program before she could read. The child taught herself how to operate the rather intuitive program. Not only did the computer provide the child with a valuable learning process, but the parents totally depended on the child when they wanted to use the program she knew. In another case, we found a young boy who taught his parents how to use PowerPoint. The examples are endless. A child may be able to understand a concept differently from the parent, and the role reversal of the child-as-teacher can be an important way the computer and the Internet can enhance a child's self-esteem. The child and parent alike may feel positively toward themselves when they are able to learn new things, master a program, win an Internet game. Their self-confidence may increase when others seek their advice, for example, because they are the family computer experts.

Enabling New Opportunities for Interpersonal Relationships

Although parents may find it difficult to enter certain areas of their children's lives, the computer gives them an excellent opportunity for common ground. The family may join user groups together, investigate topics of interest together via the Internet, or create family activity archives of graphics. Parents of children with disabilities can find many educational resources to work with their children (Bull, Winterowd, & Kimball, 1999). Parents may use the computer as a vehicle for working and playing with the friends of their teenagers. In one case, a user needed help with a computer and so he took it to a local store for service. After several days and several inquiries, the store representative said: "We're waiting for a professor in the university music department to return from a trip. He's the local expert on this problem, and we think he will be able to give us the answer." While many people learn about computers and the Internet through formal study, some of the best experts teach themselves. Often these learners turn to others for help in informal tutorial lessons or quick consults. We are grateful to so many students for what they have taught us about using the computer and the Internet. Thus, the computer provides an opportunity for links, for bonds between people, which might not happen otherwise.

Online Relationships

One woman from Korea told us that communication on the Internet is quite different in Korea than in the United States. Because Korea is a relatively small region geographically, it is easy for people to get together if they meet online. And in fact, many couples meet online as a prelude to a face-to-face encounter (Parks & Floyd, 1996). The vast size of the United States makes face-to-face connection much more difficult unless the people are meeting in a geographically specific discussion group. Make no mistake, many virtual communities do arrange to meet face-to-face. But with or without face-to-face meetings, many people are successful in creating interpersonal relationships through the Internet.

Whether or not the size of a nation has an effect on how people use the Internet, there is reason to believe that cultural differences are likely to influence perceptions of online communication (Fouser, 2001; Nomura Research Institute, 1998). Based on data collected in the late 1990s in the United States, Japan, Korea, and Singapore, it would appear that there are culturally based differences in how people view communication on the Internet. For instance, the Japanese present a more negative view of CMC compared to the Koreans. Fouser (2001) speculates that the reason for this is due to the Japanese fear that an individualist medium such as the Internet will be destructive to their traditional ways of interacting, which are centered on in-group solidarity and not communication with strangers or between individuals. Moreover, some Japanese newspaper articles, according to Fouser, have speculated that CMC available to Japanese women will tempt them into cheating on their spouses. According to the Nomura survey, the Japanese were far more skeptical about the value of the Internet in human communication than were the Americans or the Koreans. The anonymity of expression seemed to bother the Japanese, who feared that it would instigate rudeness, while the anonymity of the Internet seemed to inspire enthusiasm for the Internet among the Koreans, believing that it would facilitate free expression (Fouser, 2001).

In support of what our Korean informant told us about people meeting online and then meeting face-to-face in Korea, Fouser (2001) reports that for the Koreans, "CMC augments, rather than supplants, face-to-face communication." (p. 272).

Fouser (2001), however, recognizes the possibility of an alternative interpretation of the data: Namely, that the Japanese have not embraced the Internet as enthusiastically as the Koreans or the Americans, an alternative to the cultural theory. He calls this the computer-literacy theory, which is really a focus on practical matters and matters of convenience. Under the computer-literacy theory he points to the difficulty of inputting the Japanese language into the computer via keyboard compared to other languages, including English and Korean, simply due to the structure of the written language; he points to cost of online time in the various countries compared; and the cost of competing communication technologies in various countries. Cost, convenience, and spontaneity essentially sum up the factors of the computer-literacy theory. It is likely that these factors, as well as how

CMC works with existing patterns of communication, go a long way toward accounting for adoption of Internet communication. We could toss in the way other media in these countries portray the Internet, such as newspaper articles, books, and films (Fouser, 2001, pp. 266–267).

You may recall seeing the American film *You've Got Mail,* which no doubt helped to inform people of the way that individuals might meet online and transform their online communication into a face-to-face romantic relationship. By the way, a search of the Internet to find information on the Warner Bros. film, revealed a host of web sites, free screen saver software downloads, sound files, quotes from the film, reviews, fan web sites, and much more, focused on *You've Got Mail.* While we do not take up this topic in this book, the way in which the media influence one another and cross fertilize is quickly revealed in this online adventure in searching for information on a film, and one must pause to wonder about the ownership and hence economic backdrop to such cross fertilization. And the United States is not the only place where films have been about meeting online; Fouser (2001) discusses *haru* in Japan and *chopsok* in Korea, both about chatroom romances. Let us return to the use of the Internet for interpersonal communication.

In many parts of the world, people use computers to connect, to find lovers and friends. Computers are a medium of communication, not simply a tool for writing or computing. As such, we pay attention to the special ways in which this medium influences communication in cyberspace as an instrument of *relational communication.* Relational communication is interaction for the purposes of creating, maintaining, or ending an interpersonal relationship. These relationships may or may not involve cybersex. Although many people say cybersex is innocent fun, cybersex can go beyond pure fantasy when individuals disclose sexual information, exchange nude photos, or use the Internet to prey on others (Lambert, 2003).

One of the most interesting topics of online communication for students is online dating and relationships. Some people use the Internet to find loving partners. The prevalence of cyber-romance is reflected in the popularity of the theme in current literature, which frequently shows the conflict of poor or inadequate face-to-face communication between the partners (Janney, 2000). One woman, for example, began an online relationship that resulted in the end of her marriage. When the online relationship ended, the woman continued to make platonic and romantic relationships through the Internet in addition to her face-to-face relationships. The woman believes "the Internet has allowed her to avoid unhappiness and depression" (Biggs, 2000). One of the problems with the research in this area is that the ability to determine the truth is difficult. Some people suggest: "Assume everyone you meet online is lying." But how do you know when someone is lying? Although there are ways to check up on people online—called "googling"—you may have trouble determining the truth of the individual you encounter. With regard to lying and telling the truth, Whitty and Gavin (2001) found that both men and women lied, but for different reasons. Women lied for safety reasons—to protect themselves from men who might want to take advantage of them. They held back on giving out personal information that would identify them or locate them. Men, on the other hand, lied to disguise their true identity so that they could take greater risks with emotions and openness (p. 629). They could be less inhibited. Overall, Whitty and Gavin (2001) reported that the same characteristics people seek in f2f relationships are important in online relationships: trust, honesty, and commitment. They interpreted their interview data, however, to show that the cues that are used to assign meaning online are different from f2f relationships.

Turkle (1995) discusses yet another side to lying, one that complicates the simple dichotomy of lying versus telling the truth. She talks about people online who develop a cyberself, an online persona that is not exactly who they are in f2f life, and yet they are not lying, at least not as they see it. In fact, they might even feel more like themselves online than offline (p. 179). Turkle (1995) writes:

> The Internet has become a significant social laboratory for experimenting with the constructions and reconstructions of self that characterize postmodern life. In its virtual reality, we self-fashion and self-create. What kinds of personae do we make? What relation do these have to what we have traditionally thought of as the "whole" person? Are they experienced as an expanded self or as separate from the self? Do our real-life selves learn lessons from our virtual personae? Are these virtual personae fragments of a coherent real-life personality? How do they communicate with one another? Why are we doing this? Is this a shallow game, a giant waste of time? Is it an expression of an identity crisis of the sort we traditionally associate with adolescence? Or are we watching the slow emergence of a new, more multiple style of thinking about the mind? (p. 180)

One of the major oppositions or paradoxes of communication online is that while some people find their love, a soul mate online, others find the nightmare of a lifetime. The fact that laws have not kept up with online technology, the ease that predators have in seducing youth (e.g., Lambert, 2001; Slater, 2002), identity theft (Werde, 2003), and other online horrors underscore the need for approaching online relationships carefully (Hewitt, Biddle, Simmons, & Tharp, 2002). You may even want to consider buying and installing software that allows you to track people with whom you chat, instant message, or e-mail, to make sure you are safe (Soat, 2003).

One divorced man used the Internet in his search to find a new wife. He met a woman from Russia, and they hit it off, so he traveled to Russia to meet her. The man fell in love with the woman and her daughter, and they decided to marry. The man made all the financial, marriage, and immigration arrangements. When his beloved arrived in the United States, however, she cleaned out his bank accounts and disappeared. The man is broken hearted and broken financially. This example is just one of many of the negative results of online relationships. Do you think there are more scams in online relationships than traditional relationships? Do you think con artists can work more easily online than face-to-face? We all like to think we are good judges of people online and off line, so most people enter online relationship with a sense of trust and optimism. Like any relationships, online relationships can fail and they can work. One study found that out of 83 online relationships studied, 49 percent reported being unsuccessful (Wildermuth, 2001).

The mass communication concept of parasocial interaction may provide some explanation for the way individuals conceptualize online relationships. Parasocial interaction generally applies to situations where media audiences come to perceive certain media personalities as friends (Horton & Wohl, 1956). The person who watches Jay Leno every night, for example, may think she or he knows the late night television star, and may perceive Leno as a friend. The idea with a parasocial relationship is that through media contact, the person comes to believe she or he knows the media personality as though the media personality were someone with whom the individual has an interpersonal relationship (e.g., Rubin & McHugh, 1987). There is a similar kind of parasocial interaction in online relationships. Through limited information and text-based interaction, a person may think they know the online friend well. Much of the relationship is created in the mind, so the relationship is more imaginary than most traditional relationships. Some researchers have speculated on the ability of the participants to construct and project a self online (Turkle, 1995). Wildermuth (2001) wrote: "In the online environment, individuals are able to carefully pick and choose the aspects of themselves that they want to present" (p. 93). Moreover, Wildermuth (2001) proposed that the reduced information online relational partners have about each other can produce an intensification of the experience (p. 92).

There is no clear cycle of Internet relationship development. In terms of control of information about the self, Wildermuth (2001) suggested "that the typical development for relationships online seems to be a slow advance through more and more personal mediums where one has less and less control over the information revealed about the self (Sproull & Kiesler, 1986)" (p. 91). Control over relational distance may be one real attraction to many online romances (Sveningson, 2000, cited in Wildermuth, 2001). Baron (2000) has pointed out that "most users see e-mail as a medium that protects their private space far more than the telephone" (p. 232). In turn, the privacy shield seems to contribute to an increased willingness for people to say what they think, to express judgments, feelings, and concerns to a greater degree than face-to-face. Whitty and Gavin (2001) also reported a pattern to the stages of progression from online to offline communication. As couples developed their relationship, they tended to move from an online chat room, to e-mail, to telephone, to face-to-face encounters. Whitty and Gavin pointed out that this progression correlated with a progression in trust as well as implying a commitment to the relationship (p. 627). Ironically, they suggest that the lack of nonverbal cues or social context cues actually contribute to the ability of men especially to form relationships online. The men reported less self consciousness and less of a feeling of being judged. Anonymity may play a role here. Baron (2000) described this paradox as "The less we disclose of our physical being to our interlocutor, the more likely we are to speak our minds" (p. 233).

A person may move quickly between relational stages, go out of order, and repeat stages. But, there seem to be stages of development in Internet relationships (see Figure 8.1).

Individuals begin, create, maintain, and end their relationships in very individualistic ways, immaterial of whether they use the Internet. But these stages are presented to give you an idea of how the Internet may shape relational development.

Reflection

Have you met with success when you met someone online? Do you depend on Internet communication to reinforce quality communication in your relationships? What paradoxes can develop when people communicate interpersonally online (see Table 8.1)?

When considering the paradoxes of your interpersonal communication via the Internet, you might want to ask yourself: "Why do I use the Internet?" One study found four dimensions of computer use: (a) use because of enthusiasm of "cutting-edge" technology, (b) use for entertainment, (c) use for work efficiency, and (d) use for communica-

FIGURE 8.1 *Stages of an Internet Relationship*

—Stage 1—*Curiosity.* An individual explores, searches, talks to people in chatrooms, converses in private chat rooms, and talks via e-mail or other e-groups. The person may look on introduction services or personals webboards. There are even catalogs of women available, for example. The Internet makes the person feel connected to the world. There's a curiosity about having a real possibility of finding someone among the 6 billion people on the planet, someone who is truly in a different place, and different time than the everyday, mundane, here and now.

—Stage 2—*Investigation* is when a person finds another person through a discussion group, mailing list, referral from another person, or some other Internet means. They find out about each other. This stage is exciting for participants. The people begin telling their stories.

—Stage 3—*Testing* is when we bring up a wide number of topics. The participants seek common ground, and interpret topics within their individual contexts.

—Stage 4—*Increasing frequency of contact.* This is the time when participants increase the depth or breadth of subject discovery and discussion. They may correspond by e-mail daily or more, and may begin to use other forms of communication. The individual becomes a cognitive creation, which may or may not be grounded in reality.

—Stage 5—*Anticipation.* As the cognitive intimacy develops, the people decide they want to meet face-to-face to see if there is a physical connection. They are filled with expectations, questions about whether meeting will ruin the relationship, questions about whether each will measure up.

—Stage 6—*Fantasy integration.* Because so much of meaning is nonverbally communicated, and Internet relationships lack nonverbals, the people involved create a fantasy of who the other person appears to be, how she or he looks, acts, thinks, and behaves. This may or may not be relevant to the reality of the person. Where most people are perceiving failed expectations in f2f communication, Internet relationships may be filling gaps to make the relationship appear more integrated than it really is.

—Stage 7—*Face-to-face meeting.* The sensory deprevation during the relationship development may make the meeting intense. Fantasy and reality are different. Can each find the person she or he knows in the person they see face-to-face?

—Stage 8—*Reconfiguration.* Two people meet and realize they must bring the fantasy in line with reality to maintain, or they realize they cannot or do not want to merge fantasy and reality. There may be an immediate breakup. The pair may experience individualizing, avoiding, stagnating, or turmoil and will try to reconfigure the relationship. The physical meeting makes it difficult to return to the only Internet mode. If separating permanently, they probably part immediately after face-to-face meeting.

—Stage 9—*Already separated.* If maintaining, the couple may try to find other modes of communication to maintain a long-distance relationship. They may try to create more relational depth or breadth, while integrating a physical or proximic relationship.

—Stage 10—*Long-term relationship.* Negotiated long-distance relationship or reconfiguration into a face-to-face relationship to create a long-term relationship.

tion (Panero, Lane, & Napier, 1997). In the interpersonal context, people probably use the Internet because of their interest in the medium, as entertainment, for communication, and to meet people. And, the Internet can enable an interplay between personal and work relationships.

Many online discussion groups provide members with information, but they also can provide participants with a sense of community or belonging (Baym, 2000). One ad-

TABLE 8.1 *Paradoxes of Online Interpersonal Communication*

Available any time.	Lacks immediacy of face-to-face interaction.
Enhances relationships.	Can damage current relationships because Internet distracts from interactions.
Provides opportunities for new interpersonal relationships.	People may conceptualize the new person in idealized ways and fail to develop realistic relationships.
Transcends time and space boundaries.	Can feel shallow and artificial.
Verbal text takes some of the emotion out of relating.	Lack of nonverbals increases misinterpretation.
Offers more flexibility in the way people interact.	Some people come to prefer working with the computer than another individual.
Creates sense of belonging.	Makes people feel isolated.

vantage of groups is that people who may feel marginalized in our society—e.g., women, minorities, and people with disabilities—may be able to connect with people online in unique ways. For example, one advantage of computer group memberships is that gay men and lesbians who might be marginalized in other contexts can belong to groups that become a part of their identity (McKenna & Bargh, 1998). The relatively anonymous nature of online groups gives participants a place for interaction, which they might not have otherwise.

When participating in online groups, you may have observed that some have explicit rules of behavior and others have implied expectations. Online groups function with the same kinds of social rules that one encounters in any group. Members who participate successfully can gain greater self-acceptance and learn skills that can enhance their other social interactions (McKenna & Bargh, 1998).

Computer systems, sociotechnical factors, time, hardware, and software shape the way people communicate individually on the Internet. If you communicate with a friend via AOL Instant Messaging, for example, you can interact differently than on a WebCT or Blackboard, etc., message board. There may be differences, for example, in how the communication appears visually, time synchronization, interactivity, and privacy. In instant messaging you may not give the same kind of thought and effort you give in preparing your message on WebCT or Blackboard. The IM will be brief, may have no punctuation, use nonstandard capitalization, and show greater informality. Your IM may take on a rhythm of banter or interplay back and forth. The message posting in WebCT may take a more one-way, longer, reflective, and more formal style. This messageboard style is probably still shorter and more casual than if you wrote a snail mail (U.S. Postal Service) letter (Gurak, 2001). Thus, not only does your individual perception affect your communication, but the technology that you use may affect the way you communicate on the Internet.

Whether you are an avid Internet user or a novice, your individuality affects how you use the Internet. And your particular Internet use may affect other elements of your social interaction. Sometimes heavy users are made to feel defensive, as if they are being antisocial. But Internet users are probably at least as social as anyone else (Robinson, Kestnbaum, Neustadtl, & Alvarez, 2000). In fact, Internet users "read more literature, attended more arts events, went to more movies, and watched and played more sports than comparable nonusers" (Robinson & Kestnbaum, 1999). Paradoxically, while a problem for some people, for the average person, the Internet enhances relationships and cognitive and social interplay. What do you think? Does the Internet enhance your personal relationships? What is your communication through the Internet?

Case for Discussion

In your course discussion group, give your analysis and perceptions of the following.

We interviewed a woman who met her husband online. After months of talking, they met face-to-face. The romance blossomed into a long-distance relationship for several more months until they married. They were only married a few months when they began experiencing difficulties. Shortly after their marriage, the husband resumed his Internet conversations with other women. The wife was jealous of the frequency and nature of her husband's talk to women he met online.

In another case, a woman said that after she met her partner online, they continued to talk to each other online. In fact, they sometimes used the Internet to talk upstairs/downstairs in their home and even to initiate romance.

Relationships are a challenge, no matter how they originate. They require commitment and creativity to succeed. That seems to hold true for relationships that begin or are maintained by the Internet.

1. What are the pros and cons of meeting a romantic partner online?
2. How can the Internet be used to enhance face-to-face relationships?
3. How can Internet relationships be transformed into effective face-to-face relationships?
4. What is your analysis of the two situations presented above?

Internet Investigation

1. *Mediated Interpersonal Communication.* Prepare an essay, speech, or posting for your course discussion group which answers the following question: How do mediated interpersonal communication and face-to-face relationships compare?
2. *Nonverbal Cues.* According to Walther (1993), the lack of nonverbal cues in online relationships may depersonalize the communication causing slower relational development and differences in interaction. Do you agree? Make a list of all the ways that you think the Internet depersonalizes communication. Can you think of ways that the communication on the Internet makes communication *more* personal? How can you

convey nonverbal information online? Write an essay on the topic and post it to your online discussion group.
3. *Intrapersonal-Interpersonal Link.* What are some ways that intrapersonal communication enters into and influences interpersonal communication online? Write an essay on the topic and post it to your online discussion group.
4. *Cultural Influences.* Rice and Love (1987) were early investigators who observed that people who communicate via computers are inclined to be emotionally supportive. The socioemotional content on the Internet is similar to the socioemotional content of face-to-face communication. In this chapter, we have focused on how the computer shapes communication between individual people. Investigate how the computer shapes communication between people of different nations and post your findings to your online discussion group. Does socioemotional content transpire on the Internet between people of different cultures? In what ways does the Internet enhance or interfere with socioemotional content?
5. *Mass Communication View of Parasocial Relationships.* As you consider this information, how might you see the ideas through the eyes of an expert in mass communication? Consider, for example, the nature of parasocial interaction, a concept from mass communication study. The idea is that through mass media, people come to believe they have a type of social relationship with media personalities. When viewers refer to Dave or Oprah meaning David Letterman or Oprah Winfrey, they have some of the feelings normally attached to friendship. Investigate the concept of parasocial interaction and apply the findings to relationships on the Internet. You might want to consider, for example, work on soap opera online communities (Baym, 2000). How does the mediated nature of the Internet change the way people perceive each other on the Internet? Does the mediated nature of the Internet lead people to believe they are friends, when indeed they are not? If so, what causes this misperception?
6. *Technology of the Internet.* Although the ideas in this book are presented from the point of view of interpersonal communication, you may want to consider the ideas from a technical framework. For example, in what ways do technical aspects of the Internet influence the way individuals communicate? How are cutting edge technical advances affecting the paradoxes of communication via the Internet? How have recent military events affected technology and technology's effect on human communication?
7. Disinhibition. Disinhibition is the communication style people use online because they feel freer to disclose information, argue, and speak online in ways that are more direct than in face-to-face situations (Reid, 1991). Investigate disinhibition behavior. Begin your investigation of disinhibition online with web sites such as http://www.netsafe.org.nz/resources/resources_disinhibition.asp or http://www.irchelp.org/irchelp/misc/electropolis.html
8. *Postcyberdisclosure Panic (PCDP).* Have you ever disclosed personal information online, written a heated e-mail, or sent a confidential e-mail to the wrong person, then panicked over the potential embarrassment? Postcyberdisclosure Panic (PCDP) is that anxiety we feel when we reveal something online, then nervously await the consequences (Whittle, 1997). Conduct interviews with Internet users and ask them

the causes and nature of anxiety they experienced online. What relevant conclusions can you draw about communication on the Internet?
9. *Types of Virtual Communities.* Investigate different types of virtual communities. You might look at fans groups, who come together to discuss certain sports teams, stars, film genres, or programs. You could investigate media communities, which are sponsored by a media group, such as MSNBC.com. You could investigate people who are linked by Web Rings. Find a focus for your investigation, then see what you can learn. Share your conclusions with other people in the course.
10. *Lonely Hearts.* Investigate ways people who are lonely find love online. You could look at personal ads from prison inmates, brides available online, "penpals," or other evidence of Internet use by people who are desperate for companionship. What are the likelihood that these people will meet with disappointment or worse?

Concepts for Analysis

We have provided concepts relevant to this chapter so that you can check your understanding. These concepts are worth dissecting in more detail. As a research option, examine one concept. Use our citations as a starting point for your study. We recommend that you search for refereed articles in an online database. Find at least three solid references you can read. Summarize, scrutinize, and present your analysis to other people in your course.

Questions to stimulate your inquiry: Do you agree with the ideas as we presented them? What other points of view should be considered? What details need to be added to fully understand the concept? What can you contribute through your personal analysis of the concept? What more do scholars need to know about the topic?

Community is the ability of people to come together, to have a sense of sharing and commonality, to feel a sense of gathering or oneness.

E-mail is electronic mail or messages sent and received via the Internet, which is often used as a form of interpersonal communication.

Face-to-face communication (f2f or FtF) is a meeting in real same time and place.

Flaming is the hostile expression of strong emotions and feelings.

To *Google* is to investigate a potential date or interpersonal meeting through Google.com or other search engine to find out information before the face-to-face meeting.

Immediacy is a feeling of closeness, of sensing an emotional proximity.

Interpersonal communication is one-to-one communication, and some communication scholars consider the term synonymous with interpersonal relationships.

Reconfiguration is the stage of relationship development when two people who met online meet face-to-face, and realize they must bring the fantasy in line with reality.

Regard is to show respect for, concern; to take into account; to relate to; pay attention. Regard is used to refer to a message that acknowledges you and the feeling that gives the recipient.

Relational communication is interaction for the purposes of creating, maintaining, or ending an interpersonal relationship.

9

Groups

Debbie is a 65-year-old divorcee who lives alone in a rural area. Debbie spends several hours a day "talking" to her friends on five e-mail discussion lists. She connects to others via the Internet. Debbie's Internet use is just one example of the many ways people use the Internet to connect to other people.

When communication scholars talk about groups, they usually mean small groups of about 3 to 30 people. But the Internet redefines groups because it can provide a sense of small group interaction, while actually connecting hundreds, thousands, even millions of people. Years ago, McLuhan (1964) hypothesized that media would connect people around the planet to create a sense of community. No single medium has been able to accomplish this connection in the way that the convergence of computer technology, graphics, video, and sound does through the Internet. In some ways, the Internet has freed people of time and space conventions. The now for a person may be the yesterday of another and the future of someone else. The Internet has made the global village concept a reality.

Focus

```
TO: Students

FROM: paradoxdoc@hotmail.com

RE: Overview: Groups

What happens in online discussion groups? Lenny and
Joan tell you about how people use the Internet as a
mediator of human communication. Internet groups connect
people who have needed information or resources, for
example. Internet groups empower the self through
```

support information and self expression. Internet groups can improve the quality of life. Lenny and Joan ask you to contemplate the importance of storytelling in e-groups.

What are your experiences in chat rooms, instant messaging, electronic discussion, electronic mail, course environments, listservs, webboards, MUDS or MOOS? Lenny and Joan ask you to ponder online private language and private stories online. Perhaps the most important effect of online groups is when groups create a sense of community. Many online discussion groups provide members with a sense of support, belonging, and connection.

The nature of online groups or populations can be positive and uplifting or negative and degrading. But when Lenny and Joan think of "community," they assume groups of support and togetherness, which can be found online.

Lenny and Joan ask you to investigate characteristics of online group, including unique language, ignoring, lurking, and conversation lulls. Lenny and Joan give skill suggestions for dealing with high volume e-mail, which may be a characteristic of being enrolled in this course! They also give advice for moderating an online discussion group.

Lenny and Joan suggest you be concerned about the concepts of Internet anonymity, virtual reality, and lack of nonverbals online. Finally, they discuss health care empowerment through groups.

We will consider Internet groups that provide connection, community, discussion, support, and dating. At the end of the chapter, you should be able to answer the following questions:

1. How does the Internet mediate human communication?
2. What is the nature of storytelling in online groups?
3. How do people develop community through online groups?
4. What are characteristics typical of Online Group Discussion?
5. What are important considerations for moderating effective e-group discussion?
6. How can individuals use e-groups for health care empowerment?
7. What paradoxes have emerged because of group communication through the Internet?

Mediator of Human Communication

There are specific computer and Internet applications that appear particularly important to the discipline of communication. Many of those applications take on a broad social meaning. Of course, communication becomes the prime Internet function in such cases because the home, school, and office computers are connected to the world. Learning, health, and activist groups enable online discussants to enhance their personal growth. Information is accessed by linking up and downloading. Some of the social issues that this connection raise include:

(a) Will the Internet change societal structure by permitting a more rural society?
(b) Will the Internet enable people to be electronic hermits?
(c) Will the Internet allow for instantaneous political decision making?
(d) Will the Internet create more or less individuality, depending on how it is used?
(e) Will a cash-less society become a reality through Internet connected smart phones?
(f) Will the Big-Brother-is-watching-us fear become a reality as the Internet takes away personal privacy?

The impact of information technology touches many aspects of human communication. The Internet can handle incredible amounts of Information and enormous numbers of people. We're told our economy is now based on a renewable and self-generating resource, information, and that knowledge is doubling in less than the span of two years. The magnitude and complexity of the Internet creates a kind of magical quality about the computer-mediated communication. The Internet is attractive and ominous at the same time. The mystery and magic become more fanciful than any Magical Kingdom in Florida by providing the personalized "Magical World of the Internet." And part of the magic is that the Internet can connect anything and anybody. How does the Internet intensify connections?

Connecting People Who Have Needed Information or Resources

Anne wanted to work in genetic engineering to save endangered animal species. With such narrow interests, she wasn't exactly sure how to accomplish that goal. In one college course, however, she had done extensive reading on the subject. Anne decided to find the authors of what she considered the best research and contact them via e-mail. Through some searching, Anne came up with every author's e-mail address. She constructed a letter explaining who she was, what she wanted for a career, and included a brief resume. Anne asked each person for advice about how to learn what she needed to learn so that she could do the kind of work they do. Some of the researchers ignored her e-mail. Some gave her suggestions about the best graduate programs. But one called her, gave her a telephone interview, and hired her! This example of access is just one way the Internet can connect people.

One of the areas where the differences in information creates a major chasm is in health care. The physician has a completely different type of information than the patient. This gap can be narrowed by the Internet. By offering ill patients easy access to important health information, by providing an emotional support group to help survivors overcome grief, by giving a shy person a chance to express himself in an online discussion group about his hobby, people can use the Internet to connect to new ideas, experiences, people, and information.

Empowering the Self through Support Information and Self-Expression

The nature of groups in face-to-face situations versus online have similarities and differences (Arrow, 1997). The computer can serve an extension of each individual. Many computer users have a type of bond with their computer, which they don't feel toward other objects they own. They have this bond because when they are working or playing on their computer, they have a difficult time telling where the self ends and the computer's functions begin. The computer feels like it is stretching individuality or personhood. The computer, then, may represent each woman or each man who uses it. But the Internet ends the individualism of the computer by connecting to larger computer systems.

Through discussion groups and information access, an individual can express the self, converse in support groups, and access personal and professional advice. Group members can do so with a sense of privacy, perceiving they are engaged in a dynamic of the self. But in this private communication, the Internet allows the self to transcend traditional boundaries to engage a rich array of external connections. We have found some interesting alliances between women online, for example, and a dedication to working with and helping online "systers" (Camp, 1996).

For the self, Internet interaction is a process that goes beyond intrapersonal communication, yet is something different from interpersonal, public, or mass communication. Thus, the self has a connection outside the self in a way that is different from other types of communication. The Internet allows a more immediate or simultaneous connection than writing traditional correspondence. The Internet transfigures intrapersonal communication so that the self connects to others. Individuals sit alone at their computers, imagine the people in a discussion group, and create mental pictures and sounds of the people with whom they interact. So reading text on the Internet is like reading a novel, where the individual creates personalized sensory data. The Internet allows a more individual approach to mass communication. When watching an advertisement for a child's college fund service on television, the individual knows the message is going to millions of people at the same time. When going to the web site where the person sets up a child's college fund, the server collects data and personalizes the interaction. The same information presented in the television ad can be provided, but the message is geared to the specific Internet user, using the person's name and personal information. The web site may be communicating to thousands of people simultaneously, but it seems like one-on-one communication because the computer program can provide far more personalized data than a television ad or informational program. Thus, the Internet enables a lively interaction for the self that enables the individual to expand, create, energize, resonate, and self-reflect in new ways. Now, instead of seeing the computer as a model of the human being, one can visualize a mediator of human activity as the metaphor for the Internet (Bodker, 1997).

In addition, the stimuli on the Internet, whether they be e-mail messages or web pages or even videos, often can remain before the user for as long as they would like. In other words, they can self-pace the interaction. The user has all the time they desire to comprehend, transform, think about, and imagine. Of course, there are exceptions, such as with live, synchronous discussion, but even here there is often the option to review an archive. It may be an awkward analogy, but it is as if the Internet allows for stop motion photography of a part of the process of face-to-face communication.

Improving the Quality of Life

Smart phones offer a convergence of current technology. A smart phone is a cell phone that combines complex computer functions. For example, using a cell phone available for under $200, an individual can access a to-do list, calendar, and the Internet on their color screen. In addition, there are experiments with smart phones designed to work other computer systems, such as a computer system in the home. That way, a person can fill the Jacuzzi, turn up the air conditioning, and unlock the door to their house while driving home. By having functions connected to credit cards, for example, an individual can use their smart phone to get a Coke from a soft-drink machine. In Finland, where wireless Internet phones are popular, research suggests that through the proliferation of use, young men and women actually communicate more and feel better connected with each other. In 2001, smart phones accounted for .3 percent of world wide sales of mobile phones, but as the technology develops, sales will continue to increase (Crockett, 2001). The blending of fact and fantasy on the Internet can be a way of coming to a common understanding and uniting to deal with problems, tragedies, and disasters (Harmon, 9/23/01).

Storytelling in Groups

The Internet is a storytelling medium. Internet groups allow people to tell their stories over and over until they create their own reality. The Internet is still primarily text based, in that people share their real and imaginary stories online. Bormann (1980) suggested that by sharing stories, individuals create a group fantasy, which describes their group culture. These stories, the shared text and language choices, suggest intensification of fantasy and play.

We consider it no coincidence that people talk about "interactive" computer programs. You can gain some insights into human communication on the Internet through the concept of *interaction involvement* (e.g., Cegala, Savage, Brunner, & Conrad, 1982). Cegala (1984) suggested that people with high interaction involvement could better process information, recall details, and had stronger egos. Can the same be said of highly involved Internet users? Cegala said that low involvement people feel uncertainty about interaction with strangers. Can we deduce that people who use the Internet to meet strangers are high involvement people? We may be able to gain insights into meaning by looking at the redundancy of interaction on the Internet as participants tell and retell stories.

Chat room talk and e-group conversations may provide light-hearted interplay between users. Online Internet electronic discussion is a specific form of communication (Waldeck, Kearney & Plax, 2001). A *chat room* is a forum for synchronous discussion. Usually each message uses only a few words, which includes abbreviations. Chat rooms are generally open to the public or all members of a group. Some services limit participants to 6 people, others limit them to 30 participants. *Instant messaging* is when two individuals send messages via the Internet, which are synchronous. Sometimes called chat or IM, instant messaging is usually in a private place and may involve several people at once. *Electronic discussion* is a more generalized term which refers to online groups or lists that communicate synchronously or asynchronously via electronic mail, course environments, listservs, or webboards.

In chat, participants give brief, often one-line statements, but in e-mail discussion, users tend to develop their ideas in a paragraph or more. Chat is synchronous, in that all parties can talk at the same time, although e-mail discussion is not necessarily synchronous. Although useful in one-on-one interaction, chat and instant messaging can be somewhat confusing, particularly with group interplay of more than a few people. The speed and confusion of chat can create a sense of play. Sometimes through chat or instant messaging, the communicators create an expressive connection. They develop a quick banter back and forth which is almost a type of language game. In these cases the purpose of IM may not be so much for the purpose of exchanging information, but of creating an exchange or expression between two people.

Perhaps you belong to an e-group or electronic discussion group operated by a listserv, through which you receive e-mails from group members. E-group or electronic discussion group or a discussion list is a gathering of people who interact together via the Internet. They often have a common bond of some kind, such as that they like to discuss politics, sex, or a hobby, or provide support regarding a particular problem. Another type of e-group is a discussion board or webboard, which is a service set up so that electronic communication is conducted on a server. Many groups are set up to communicate through the webboard method. Through a *webboard,* a person uses a name and password to enter into the system. The advantage of a webboard is the technical convergence available in one place. The group's e-mail, synchronous chat, graphics, streaming video, and web pages can be located in one place. The webboard usually delineates conversation by threads, so the members can more easily focus on just those conversational threads of interest.

When going on an e-discussion list, most lists require a name attached to the subscription. Some listservs allow the name "anonymous," in which case no name is provided to the list moderator or the list members. If the person posts to the list, only the e-mail name exists. Many people believe they can have anonymity this way. Of course, anyone with good computer skills can track down an individual's identity because the computer leaves an information trail. The average person on a list, however, does not know how to track or identify another person (Poole & Hansen, 2001).

In communication, which is the primary activity, knowing the identity of those with whom you communicate is essential for understanding and evaluating an interaction. Yet in the disembodied world of the virtual community, identity is also ambiguous. . . . In the physical world there is an inherent unity to the self, for the body provides a compelling and convenient definition of identity. The norm is: one body, one identity. (Donath, 1999)

In contrast to this norm, Internet discussion groups allow an individual to play with identity and to be denied access to the identity of others. Thus, identity is a cognitive creation.

Some people subscribe to multiple lists. Others rotate from list to list looking for interesting conversation or a sense of belonging. For some people there is no commitment to a particular group, but an interplay of the initial interaction stage of relationships. People enjoy the interplay of meeting people and telling their own stories. We have observed that most lists can be—at least for a brief period—as interesting as other forms of entertainment. We heard, for example, more than one list member talk about conversing with friends online while a spouse was in another room watching television. These people use the Internet as their entertainment medium. Consider the playful exchange between two Internet list members illustrated in Figure 9.1. Note the informal writing style, lack of punctuation, and use of capitalization.

Some lists have consistent stories, myths, and virtual interactions (Aitken, 2000). The sense is much like the play of MUDs and MOOs. On one list, for example, there was a virtual band with several members who discussed playing their instruments together. None of this interplay was face-to-face, but the people formed a musical group in their imaginations.

Consider Figure 9.2, which gives examples of private language and private stories. Although a list of 300 women, this conversation line was the dominant exchange one morning. Note the mother-child interaction. These are adult women conversing, at least they say they are. When someone on the list is happy, they start "jumping on the couch." The "list mom" tries to make the kids behave. As an example of myth, consider the snow comments. Several Northerners on the list enjoy snow for winter sports. To make it snow, one says she does a snow dance with a spoon, which some members from the lower 48 States complain causes snow and rain storms in their areas.

Although these kind of private stories may be displayed in an open forum simply for fun, they may also indicate power and status on the list, perhaps indicating who has elite status on the list as members of the *ingroup*. Some of these stories are convoluted and some are never explained. The list participant must figure it out on her own, ignore it, or leave. The stories and language give cues to position within the social group. One woman who joined the list unsubscribed angry because of the "cliques" whose conversation was unrelated to the list topic.

FIGURE 9.1 *Internet Informal Style*

xxx: yea how many lists are you up to now?? LOL last count we think you told me 40 something??
yyy: LOL gave them all up but 10 now with the new one but figure 3 are with you :)

FIGURE 9.2 *Example Dialogue Demonstrating E-group*

> **xxx:** Mom: I didn't want you to think that I was an ungrateful kid for not thanking you for my trivia prize, but I didn't receive anything :(Oh well, maybe next time.
> **yyy:** No, sweetie Mom is way behind on things & send out small piles each payday so be patience with me :)
> **xxx:** PS JUMPS ON COUCH AND SINGS JINGLE BELLS >>
> LOL Jingle Bells :) I'm ready for summer :) Oh what the heck Jingle Bells Jingle Bells Dashing though the snow on a one...... WOoohooo im signin up now LOL like I need another list eh? *GRIN*
> **yyy:** LOL Me Too but I'm addicted :) Mom
> **xxx:** NOOOO momma you gonna have xxx dancin with that dang spoon again LOL "I sings" Im just singin in the rain just singin in the rain ohhhh what a wonderful feeling im happy again LOL oops im in tornado alley most time where theres rain theres them swirly thingys LOL Ohhh momma you are doin good with that 12 step program LOL
> **yyy:** LOL I'm doing really good I gave up all those lists & smoking all at the same time LOL LOL :) Mom
> **xxx:** OMG lists AND smoking!!!! what a woman LOL
> **yyy:** I'm in Oregon so you know I hate this song you can have all our rain LOL > :)
> **xxx:** NO mom I want snow so I can throw a snow ball at my sister xxx please say snow please. Let it snow let it snow let it snow

Developing Community through Groups

Many online discussion groups provide members with a sense of support, belonging, and connection. The nature of online groups can be positive and uplifting or negative and degrading. But when we think of "community," we think of the concepts of support and togetherness, which can be found online (Baym, 2000; Chernyshenko, Miner, Baumann, & Sniezek, 2003; Jones, 1995; Schatz, 2002). Through online interaction, people relate to other people; they communicate about and with each other as they establish community (Rheingold, 1997). Remember, by community, we mean the ability of people to come together, to have a sense of sharing and commonality, to feel a sense of gathering, oneness.

Writing about online communities of fans of televised soap opera, Baym (2000) speaks of community as "organized, like all communities, through habitualized ways of acting (Hanks, 1996; Lave & Wenger, 1991). Viewed in this way, the limits and possibilities of computer networks and mass media texts are preexisting contexts that become meaningful only in the ways in which they are invoked by participants in ongoing interaction" (pp. 4–5). The key point here is that community is largely defined by how we communicate with one another, the communicative behaviors of participants—how we talk to one another. Baym is writing about soap opera news groups. News groups and other online groups, which typically form around a topic of shared interest, have been very popular.

Topics, Inc. <http://www.topica.com>, a San Francisco-based company established only in 1998, was able to secure $44.5 million in three rounds of financing from several blue-ribbon venture capital and private equity firms. Launched in early 1999, it provides a free directory of e-mail lists and free hosting for those wanting to run lists. Topics claims there are 200,000 public e-mail lists distributed to 15 million subscribers, producing 17 billion e-mail messages annually. The service currently manages about 100,000 lists.

In August 2000, Yahoo! <www.yahoo.com>, based in Santa Clara, California, bought eGroups, Inc., a San Francisco company developed under the aegis of CMGI, whose eGroups brand grew quickly to 13th place in popularity among all web sites. When eGroups was purchased, it claimed it hosted 14 million members and 600,000 active e-mail groups. eGroups claimed to be the most widely used platform for e-mail discussion group communications.

Most recently, the Mountain View, California-based Google <www.google.com> February 2001 purchase of the failing Deja.com (formerly Deja News), a directory and search engine of the Usenet news groups, might be a strategic coup. (Conhaim, 2001, para 7–9)

We have heard some people who have never participated in e-groups talk about "real" versus "virtual" relationships. The word "virtual" suggests "pretend," but we found online communities and friendships have a reality people sometimes fail to achieve face-to-face. Some people are able to create strong bonds and feel connected in social and intellectual ways. For some users, e-mail groups are places where people exhibit who they are, while they validate each other and their lives.

A distinction can be made here, at least according to one view. Some would maintain that most discussion groups are not communities because a community requires trust (Rheingold, 1993). But that sense of community may be the most important attribute of the process of e-group communication. Online communities tend to contain much self-disclosure or telling of personal information, which can be important communication for the sender and receiver. Do people disclose more information faster online? Some scholars believe so. Are e-groups the same as face-to-face groups? Online groups function with the same kinds of social rules that one encounters in any group setting. Members who participate successfully gain greater self-acceptance and learn skills that can enhance their communication in other contexts (McKenna & Bargh, 1998). What often starts as an interest in a particular topic that defines an online group, winds up as a rich interpersonal world for its participants. Baym (2000) found this in her research on a television fan group: "When people first start reading rec.arts.tv.soaps (r.a.t.s.), they are attracted primarily to the wealth of information, the diversity of perspectives, and the refreshing sophistication of the soap opera discussion. Soon, however, the group reveals itself as an interpersonally complex social world, and this becomes an important appeal in its own right. For many, fellow r.a.t.s. participants come to feel like friends" (p. 119).

Characteristics of Online Group Discussion

We have heard people on lists call the moderator the "list mom." A list moderator can set rules, intervene in online conflict, set the interaction tone. As one person said, "I'm here in Alaska, far from my family, so the people on this list is my family." For some users, the group provides community. If you belong to an e-group, you may participate for various reasons. Perhaps the most important purpose for an e-group is to make interpersonal connections, but entertainment, professional growth, power, and self-exploration also fuel discussion groups. The motivations for online discussion seem to align closely with those of interpersonal communication: affection, control, escape, inclusion, pleasure, and relaxation (Rubin, Perse, & Barbato, 1988). Kraut, Lundmark, Kiesler, Mukhopadhyay, and Scherlis (1997) analyzed reasons why people use the Internet and found the primary reason was pleasure, but creating interpersonal relationships was high on the list.

Even the research related to student e-mail uses supports the idea that students use the Internet for positive connections. Online discussion seems to have potentially positive effects for students. McCormick and McCormick (1992) found that e-mail of college students served a primarily social function, with about 25 percent of messages containing intimate content. Few messages in the study conveyed hostility or social inappropriateness. We have observed that when negative messages are sent, however, they seriously threatened the group climate and the comfort members feel about expressing themselves. Students have traditionally used e-mail to encourage one another academically and personally (McCormick and McCormick, 1992). E-mail also may have positive effects on social behaviors such as collaboration and motivation (Brandon & Hollingshead, 1999; Shedletsky & Aitken, 2002). And for the reticent or shy person, e-mail discussion offers certain advantages enabling positive connections to others (Kelly, Duran, & Zolten, 2001).

We have observed many communication characteristics exhibited by users of online discussion, including self-absorption, unique language, ignoring, lurking, conversation lulls, participation in multiple lists, experimentation with identity, virtual reality, conflict, diverse interactants, and lack of nonverbals (Aitken, 1999). Discussion groups offer an experiential library from which to learn. You can learn much by studying the archives of various e-groups. Archives are a collection of all the messages sent by members of that group, which may be accessible to the moderator, members, or the public. And public archives are usually considered fair game for researchers. Because the individual knows her or his words go into a public archive, one does not need permission to use their e-mails in research.

Further, archives make an interesting study because online groups exhibit norms and rules in ways similar to face-to-face groups, so e-groups can provide models to study group communication processes. As explained by McKenna and Bargh (1998), some e-groups have explicit rules of behavior and others have implied expectations.

Unique Language

Some groups develop their own language conventions. Often the members of online groups are able to read and speak a type of *argot* exclusive to their culture. An argot is a unique way of speaking—almost a unique language—used by members of a particular group. For example, you may have noticed that many online discussants capitalize nothing, combine words, run paragraphs together, use context-specific terms, and use abbreviations. Some of these techniques may be used to increase typing speed; others suggest exclusivity. The argot makes it easy to know who is experienced in a group and who is new. For those interested in understanding computer technology as communication, discussion groups can provide a laboratory for analyzing language, particularly when reading group archives.

Ignoring

Most group participants see their e-mails as part of a conversation, so when an individual takes the time and energy to write to a discussion list, she or he may expect, even demand, a response. An overt response does not necessarily happen and being ignored often prompts discussion on lists about what the behavior means.

Ignoring may have one of several motivations. Some groups are large volume lists—members may delete days' worth of mail or select "no mail" when they are busy—so members' time demands require that they be selective about responding.

E-groups are largely for entertainment, and when life is busy, the participants simply do other activities.

Some people may want to avoid conflict or believe the mail does not warrant the time required for a response.

If one individual has spent considerable time writing to another and is ignored, that ignoring can seem more demeaning than a negative response. We observed a case, for example, where a person we'll call "Laura" received no response to a "heartfelt" e-mail, so she ceased participation in hopes of obtaining attention—rather like running away from home. In this case, the list operator soothed Laura's feelings and enticed her to come back into dialog, at which time members said how much they liked Laura.

Whether a participant or moderator, you will want to be alert to possible misinterpretations if you choose not to respond to messages and may need to resolve or mediate conflict caused by a participant's reaction to a lack of response. The sender might think of the situation as being similar to a face-to-face situation, where the other person just gets up and walks away without comment, and feels rebuffed or insulted.

Lurking

On one list we joined, a moderator said that only about 10 percent of the people on a list participate. The rest are called "lurkers" because they read the list discussion without joining the conversation. Some lists do not tolerate lurkers and simply unsubscribe anyone who does not contribute. We have observed that certain participants will dominate e-mail discussion just like some people dominate the talk in face-to-face groups. Perhaps some quieter individuals or people from less talkative cultures are more chatty online than in face-to-face situations. To prevent lurking, however, some moderators may require a certain number of postings from each group member.

Conversation Lulls

Conversation intensity on a list comes and goes. On some lists, there seems to be little to say after the introductions. To prompt dialog, some moderators ask a question each week to help generate conversation. One strategy that helps group interaction is when group moderators post interesting information on a regular basis, simply to arouse interest and involvement. Although some posts may seem irrelevant, they can enhance the social dimension of the discussion. The problem is this kind of message can take a work group off task.

Skill Section: High-Volume E-mail

Given the enormous volume of possible e-mail from a large course environment or multiple e-lists, we asked people who participate in lists generating dozens of e-mails a day, how they process so many messages. Because list e-mail volume can be overwhelming, you may find such strategies useful. Their approaches included:

(a) Focus on certain subject lines or conversation threads.
(b) Focus on reading messages from certain individuals who are interesting.
(c) Speed read, read only the opening line, or scan the message while opening a series of messages (then returning to read only the messages of interest).
(d) Receive digest forms (the entire day or week's messages sent at once), no mail, or delete messages when busy.
(e) Maintain separate e-mail accounts for different purposes or use a e-mail program sorting system that separates e-mails according to the sender or subject line.

Internet Anonymity

To most people on lists, an anonymous e-mail seems to instantly lack credibility. The anonymous person may be screened out of conversation or negated on the list. The lack of name may seem more important to some list members than the content of her or his messages. If one lies and makes up an unusual first and last name, however, the person's message is more valued by many list members. As one participant explained: "Some people have flamed me because I won't give my name, like a name makes a difference if someone's lying." So, why might someone on a list want to be anonymous in e-discussion?

(a) The person has been harassed or stalked.
(b) The person has been embarrassed on a list and prefers to risk verbal attack to a person attached to a fictitious name rather than a real name.
(c) The person does not want to be recognized by users who are on multiple lists with each other.
(d) The person may want to make a political statement in favor of the right to privacy and against government surveillance online.
(e) The person may desire privacy because she or he is engaged in a discussion group that exchanges personal or intimate information.
(f) Of course, people join lists for all kinds of reasons, and making money is a common motivation. If a list allows people to join the list without identification, salespeople may join for the purpose of sending advertising spam.

Virtual Reality

Virtual means something that happens in electronic space, such as virtual reality, virtual groups. Virtual is real, but exists electronically. Through the Internet, people who would never come together under normal face-to-face circumstances come together online. Discussion participants have to deal with more ideas from a wider array of people than they normally meet at work or home. So, people may encounter different communication content and style on the WWW. As Rose-Neiger (personal communication) explained about communication style:

> In the past, I've used online discussion forums to enable German and U.S. students to "meet" and discuss interculturally relevant topics. I was struck by the directness of the U.S. students. Internationally, both Germans and Americans are considered quite direct. They say what they think and mean what they say. However, from an American point of view, Germans are frequently perceived as being too direct. For example, studies have shown that Americans tend to use more modal verbs in a discussion ("it could be this way"), whereas Germans tend to say ("this is the way it is"). Both are polite, of course, but they are acting under different norms. In our courses (English for Professional Purposes), we therefore encourage students to use such phrases as "You may be right, but . . ." instead of "That's wrong because . . ." or "In my opinion, xyz" instead of "XYZ!" However, in the "authentic" international encounters we created—in the online discussion forum—the Americans tended to disagree without using the politeness markers (softeners). Did this reflect the students' normal speech patterns, or are we dealing with different norms on the Internet? Are there any differences between the rules of politeness (or even civility) for face-to-face communication and for communication on the Internet? (discussion group posting, 4/22/01)[12]

Not only might communication style be different in a discussion group, but the ideas and information may seem quite unusual too. Online strangers may challenge your belief systems and threaten the values by which you live. When some members face the resulting disagreement, they may storm out and unsubscribe from the list. We have observed that some people behave in disturbingly crass, prejudicial, and hurtful ways online. We wonder if they lack responsibility because they can make derogatory comments anonymously without having to face the other person. As an e-group participant, you may need to find new ways to manage conflict and increase your tolerance of diverse ideas and communication styles.

Lack of Nonverbals

We've talked about the problems of limited nonverbals, but the principle bears repeating. You'll want to recognize the importance of the lack of nonverbal communication online, particularly during conflict. This lack of nonverbals instantly reduces the ability of people to manage impressions and persuade each other in the same ways they do in face-to-face communication. Although photos and video are changing e-discussion, without nonverbals there is no way to hug, to give a dirty look, or to handle conflict except through words. Despite using "LOL," you cannot hear an infectious laugh that releases the tension in a conflict. You may need to clarify your meaning by indicating when you are kidding, confused, or upset.

Skill Section: Considerations for Moderating Online Discussion

So far in this chapter, we have been concerned with discussion members. Here are some suggestions if you will be moderating a work, educational, or recreational online discussion.

Consider Creating a Pragmatic System in Advance

Whether you are working in a personal, academic, or business context, if you are list moderator you may want to set behavioral expectations and select an operational system for online discussion. Some list moderators believe that a mature person needs no rules of behavior. Others have elaborate rules such as: stay on topic, no flaming, show respect, use subject lines carefully. Of course, if you set rules, you will have to monitor behavior, which means extra effort. Reading and screening all e-mails in advance can be extremely time consuming. Of course there are programs available that screen for certain language. Many group participants know how to talk around those programs, by using symbols like **$ex, FU,** and **A$$,** for example. To what extent do you want to control e-discussion? If you are moderating a work discussion for a creative flow of ideas, will you use a laissez-faire approach? If you are moderating a discussion group, will you be responsible for the content of the group messages?

Although most Internet discussion systems make e-mails compatible, group moderators may need to set certain parameters to make sure that e-mails are readable and attachments, if allowed at all, contain no viruses. Further, moderators may want to prohibit hyper text markup (HTML) e-mail because it can include programming language that allows the sender to track and view private electronic e-mail without the other person's knowledge, which risks user privacy (Harmon, 2001). Moderators will want to use a secure system so that e-discussion is private and safe.

Consider Encouraging Brief E-Mails

E-mails usually are very short—often under 100 words—because they are designed for rapid scanning. Group discussants will want to respect the busy-ness of life. List members generally consider giving a speech or being long-winded in an e-mail as inconsiderate and poor form. Sometimes participants put a *long-warning* in the subject line. Better to send multiple e-mails with clear subject lines, each on a different topic, so discussants can follow a conversation thread of interest.

Consider Avoiding Private, Direct E-Mails

On discussion lists, when a person hits *reply,* the e-mail response usually goes to the entire group. Seldom do lists automatically give private responses. Individual-private replies are considered poor form in many discussion groups, although the private-reply-format is often used for low-volume announcement lists where members are discouraged from talking to the whole group online. An *announcement list* doesn't allow direct interaction between participants, but is one-way flow from the moderator.

Problems can arise when members decide to send direct, private e-mails to each other, instead of communicating to the whole list. The group audience serves as a check-and-balance system to keep individuals accountable. Receiving direct, hostile, or unwanted e-mails is like having someone yelling face-to-face in one's living room. Participants will want to respect—not violate—individual privacy and personal space. Moderators may want to quiet e-mails of violent and disturbing content by requiring subject line warnings of potentially offensive content; students should not be forced to confront disturbing e-mail content.

Consider Allowing Confidential Talk

Participants may want to make up a name so no one knows who they are on a list, and you may want to encourage members to change names of real people they discuss in their posts. Confidentiality may be crucial if the group has public archives. With confidential Internet discussion and clear markings of subjects, you'll find it easier to create a place where members may comfortably discuss—or avoid—difficult topics.

Health Care Empowerment through Groups

Today's patient has easy access to information through the WWW. In a sense, a new resource exists today: the patient. Because of the availability of information, some physicians actually schedule more appointment time when patients say they are coming to discuss information they found on the Internet (Beirne, 2000; Wright & Bell, 2003).

Whether the information is reliable, however, is a major concern. The American Medical Association, the National Cancer Institute, and others have guidelines for what is quality medical information on the Internet.

HI-Ethics, Internet Healthcare Coalition, and Health on the Net are among the groups that are trying to establish standards for assessing health information and product sites. At this point, some web sites may comply with standards. But even a compliant web site today may contain out of date, inaccurate, and unreliable information tomorrow.

One problem with using search engines to find good sites is that many search engines charge to show web sites at the top ranking. University, nonprofit, and government sites generally do not pay for the search engine positions. Small sites can grab from big sites and steal content. Much of the effect is economic-related.

Patient-Doctor Communication

One of the key communication problems between physicians and patients is the doctor's preference for technical health discussions. Physicians often fail to discuss the kinds of issues that patients want to discuss. And that communication pattern continues in e-mail discussions just like it occurs in face-to-face interactions (Bruner, 1999; Grandgenett, 2001).

> The World Wide Web and other Internet-based resources have many of the characteristics necessary for persuasive communication and may, in fact, constitute a hybrid channel that combines the positive attributes of interpersonal and mass communication. The notion that the Internet features many of the persuasive qualities of interpersonal communication makes it a prime candidate for the application of key behavioral science theories and principles to promote healthier behaviors. The broad reach that the Internet shares with many mass communication channels indicates an economy to Internet-based efforts to communicate with large audiences. It is concluded that if the Internet can be used for persuasive health communication and its reach continues to expand, it is time for public health professionals to explore the design and evaluation of Internet-based interventions directed at health behavior change. (Cassell, Jackson, & Cheuvront, 1998)

Patient Storytelling Groups

In online mutual help groups, "the most prevalent form of communication was providing self-disclosure, followed by providing information/advise." In one particular study, conflict was infrequent and communication was generally warm and supportive. (Klaw, Dearmin Huebsch, & Humphreys, 2000)

One interesting study found that people who participated in online groups cared less about what the group moderator thought of them than they did in face-to-face groups. Thus, they were more likely to give potentially embarrassing information in the online groups (Walston & Lissitz, 2000).

One idea that can apply to many contexts is tailoring CMC to the individual. This approach may work well in the health care context, for example. Medical personnel "can now provide health information and behavior change strategies that are customized based on the unique needs, interests, and concerns of different individuals." (Kreuter, Farrell, Olevitch, & Brennan, 2000).

The interactivity of a site may affect how a web site affects an individual. Nonprofit and government organizational sites that focus on diseases or medical practice, sites designed to open up dialog and information, and sites designed for alternative medicine, education, and sales were less likely to be interactive. Sites that provide choices, responsiveness, and interpersonal communication probably are more effective for the user (McMillan, 1999).

Accuracy of Information

There is no way to know exact statistics, but there seem to be many people with serious health problems who participate in e-group discussion. Although it is possible that these people are simply open about their health problems, more likely, these many people who have mobility problems find this to be an interesting way to interact with people.

As we discussed in the informatics chapter, inaccurate information is a serious problem online, and related to health problems, inaccurate information can have catastrophic effects. We have heard representatives from the National Department of Health discuss their concern about figuring out a way to check valid online health information. A recent controversy against smallpox vaccinations to protect against a terrorist attack, for example, revolved around the mistaken belief that immunizations cause autism. Online, you can find information from QuackWatch.Org *http://www.quackwatch.org/03HealthPromotion/ immu/autism.html* But for every site with valid scientific information, you can find others with false information. Some people looking for cures turn to the Internet instead of the medical community.

Reflection

How do you use the Internet to connect with other people? Do you use e-mail and webboards? How do the ideas in this chapter compare to your personal experiences online? What paradoxes have you observed in online group communication (see Table 9.1)?

Users participate in online discussion to make interpersonal connections, for entertainment, professional growth, power, and self-exploration. There are many types of groups: business, personal, learning, entertainment, hobbies. Communication characteristics exhibited by users of online discussion include self-absorption, unique language, ignoring, lurking, conversation lulls, participation in multiple lists, experimentation with identity, virtual reality, conflict, diverse interactants, and lack of nonverbals. Online discussion is more involving than face-to-face discussion, as participants can become more immersed in the communication (Coleman, Paternite, & Sherman, 1999).

Unlike face-to-face discussion, when people talk online, they are less inclined to see the other group members as individuals. Instead, they react to the messages without necessarily knowing or processing what individual sent each message. This principle is called *deindividuation theory*. But deindividuation alone does not account for some of the highly negative communication that happens online (Coleman, Paternite, & Sherman, 1999).

Certainly few of us actually understand the workings of the Internet, but we come to better understand ourselves and the Internet "as we project our sense of usefulness onto it" (Thorington, 1999, para. 3). The Internet intensifies our sense of participation and place in the world view. We fit because we communicate.

TABLE 9.1 *Paradoxes of Group Communication on the Internet*

Connects people.	Groups reject people.
Allows openness to new people.	Shuns diverse ideas and enables extremists to find people of like mind.
Activists can organize online to act.	People can "talk" online instead of taking action.
Support groups offer hope.	Support groups reinforce victimization and depression.

As you evaluate the ideas of this chapter, consider how Internet interplay intensifies a perception of common ground between people. Do you use the Internet to connect to other people literally? Figuratively? Both? Do you think Internet interplay can create a kind of global consciousness because it connects you to humanity around the world?

Case for Discussion

In your course discussion group, give your analysis and perceptions on the following interviewee comments.

"I remember planning to buy my first computer. I was working on a graduate degree, and I thought it would help to have a computer. After careful consideration, I remember saying: 'I think 126 k is all I'll ever need.'

Today I have a computer at work and two at home. One at home actually belongs to my son, but at home I can use two computers at the same time, and complain about how I don't have enough capacity with only 100,000 k. The dumb thing is always complaining about 'insufficient memory.' Well, computer, get a grip; I have the same problem. You can learn to cope.

One day I asked my husband what we would do when our children went off to college. We spend so much of our lives revolved around the kids' activities. His answer: 'We'll sit next to each other, cuddle, and parallel compute.' "

1. How is changing technology affecting your connections to other people?

Internet Investigation

1. *Examine Health Sites.* Examine health-related sites. Explain the criteria you use to determine which sites are viable and which are not. Write a brief summary of useful sites, including the URL.
2. *Health E-Group.* Examine the archive of an e-discussion group designed to help people meet each other. Find examples of how people seek common ground. How successful do you think the people are in making long-term connections?
3. *E-Group Participation.* Find a discussion list of interest and participate during this semester. Your instructor may ask you to discuss, post and e-mail, or give a report about your observations about the experience. Participate in an e-mail discussion list. What causes an emergent community?
4. *Health Information Survey.* Conduct a survey of how people use the Internet for health information. Create a list of questions. Post the results and some anecdotal comments to your course discussion list. Your instructor may add or substitute questions to your list. Survey questions:

    (a) Can you separate site information quality from aesthetics quality? If yes, how?
    (b) What kinds of sites do you, your family, and your friends access for health information?

(c) Do you and your health care providers agree on which Internet sites are high quality?
(d) When it comes to health information, can you distinguish good sites from bad sites? If yes, how?
(e) How do you act upon the health information you find on the WWW?

5. *Mass Communication View.* Reflect on the information of this chapter through the eyes of a mass communication expert. *Uses and gratification* theory suggests that people use the Internet because of the rewards they receive. Investigate uses and gratification theory and speculate how the theory applies to the groups who communicate via the Internet (e.g., Althaus & Tewksbury, 2000).

6. *Gender Differences in Online Communication.* Some researchers believe there are general differences in communication style based on a person's gender. Those communication differences may manifest themselves in online communications (Camp, 1996). Men are more likely than women to criticize and ridicule online (Herring, 1999). Men are more likely than women to represent themselves as experts and attempt to dominate the online discussion (Soukup, 1999). Investigate the research on gender differences in communication, then conduct your own analytical research in an online discussion group. Do you agree or disagree with the published research? What conclusions did you draw from your investigation?

7. *Do You Have Net Presence?* Net presence is being known or having a type of power in a discussion group or on the Internet (Agre, 1994). Individuals may, for example, participate in several online discussion communities and become known to a group of people. We remember the first time something we did in a classroom began to circulate online, and returned to us—free of our names—months later from a totally unrelated source. Do you have Net presence? How is Net presence achieved? Examine the archive of a discussion group and identify at least one person with Net presence. Explain why the person has Net presence.

Concepts for Analysis

We have provided concepts relevant to this chapter so that you can check your understanding. These concepts are worth dissecting in more detail. As a research option, examine one concept. Use our citations as a starting point for your study. We recommend that you search for refereed articles in an online database. Find at least three solid references you can read. Summarize, scrutinize, and present your analysis to other people in your course.

Questions to stimulate your inquiry: Do you agree with the ideas as we presented them? What other points of view should be considered? What details need to be added to fully understand the concept? What can you contribute through your personal analysis of the concept? What more do scholars need to know about the topic?

Anonymity is when a group participant is permitted to anonymously post messages to a group so that they cannot be identified by group members.

An *argot* is a language convention developed by an online group. Often the members of online groups are able to read and speak a type of argot exclusive to their culture.

Chat rooms are generally open to the public or all members of a group, where there is lively, quick, synchronous conversation.

Electronic discussion is a generalized term which refers to online groups or lists that communicate synchronously or asynchronously via electronic mail, course environments, listservs, or webboards.

Instant messaging is a way for two individuals to send synchronous messages via the Internet. Sometimes called IM, instant messaging is usually in a private place and may involve several people at once.

Interaction involvement theory suggests that people with high interaction involvement can better process information, recall details, and have stronger egos (e.g., Cegala, Savage, Brunner, & Conrad, 1982; Cegala, 1984).

Lurking is to read but not post messages to a discussion list. Some users consider lurkers to be similar to voyeurs.

Virtual means something that happens in electronic space, such as virtual reality, virtual groups. Virtual is real, but exists electronically.

Notes

12. Dr. Ingrid Rose-Neiger is a Professor in the Social Sciences Department of the Karlsruhe University of Applied Sciences in Karlsruhe, Germany.

Part IV

Contexts outside the Home

This section examines the contexts of Internet communication or where the communication takes place outside the home. Specifically, we discuss work and school contexts.

Chapter 10—Workplace Contexts—Gives you ideas about systems, roles, tensions, and decision-making influenced by the Internet. We discuss influential factors of organizational climate. We also explore some characteristics of e-commerce.

Chapter 11—Educational Contexts—Gives perspectives about Internet communication in schools, particularly higher education. The Internet intensifies learning. It allows anyone to learn anything, anytime, anywhere. The Internet intensifies effort and collaboration.

10

Workplace Contexts

> *When applied to the organizational setting, sequentially, or shapeshifting, represents the emergence and continual shaping of new organizational forms in response to rapidly changing environmental circumstances; simultaneity is evidenced in the balancing of contrasts such as centralization and decentralization; and the concept of sociality represents the need to maintain the known or core ideology as a mechanism to support change and the evolution of a new sense of place. Taken as a whole, the concept of protean shapeshifting suggests that change and innovation are about the simultaneous disruption of place and the seeking of a new sense of place. Shapeshifting is about the simultaneous occurrence of chaos and order. Finally, shapeshifting is about simultaneously maintaining core values while supporting continually changing practices. Lifton's work originally was applied to the notion of the self and the complex challenges for individuals. The metaphor of Protean Places extends the concept of proteanism to organizations with their changing forms and processes.*
>
> —(Shockley-Zalabak, 2002, pp. 237–238)

Focus

```
TO: Students

FROM: paradoxdoc@hotmail.com

RE: Overview: Workplace

Lenny and Joan explain that there are various elements
to business communication on the Internet. If you were
```

> to search databases, for example, you could find thousands of articles and many books about business on the Internet. You can find technology businesses, e-businesses that are accessible primarily through the Internet (also called dot com companies), and ways that most American businesses are using the Internet.
>
> Consider these ideas. The system affects the communication. Meaning is blurred because the roles and boundaries are blurred. Paradoxes abound. Decision making is influenced by the Internet. Internet play can interfere with business productivity.
>
> The Internet communication affects organizational climate. Communication on the Internet can influence structure and constraint, the sense or lack of individual responsibility, feelings of warmth and support, conflict and tolerance, identity and group loyalty, and risk.
>
> Lenny and Joan discuss online business interaction, including consumer communication, interpersonal communication, collaborative communication, groupthink, communication overload, and risky communication. Finally, in the skill section, Lenny and Joan give tips for managing communication.

In this chapter, you will examine human communication on the Internet in the business or professional context. When you finish reading, you should be able to answer the following questions.

1. What is the nature of business communication on the Internet?
2. What is the nature of e-business communication?
3. What are some research findings about business in the Internet context?
4. What paradoxes operate in the context of business communication?

Perspectives of Understanding Organizational Communication

You can approach your investigation of organizational communication from one of several standpoints (Neher, 1997; Redding, 1992; Sypher, Applegate, & Sypher, 1985). The *functional perspective* examines the effectiveness of organizations. The functionalist is concerned with an organization that functions efficiently and productively. The functional approach looks at what was intended and what happened in an organization. Laws of behavior give us a way of predicting and controlling behavior in the organization. Of particular interest is organizational operations, behavior, and predictable effects. The *critical perspective* is primarily concerned with power and control in organizations. The critic is more concerned about why the organization is running effectively. Who benefits? They may be concerned with material or humanist outcomes.

The *interpretive perspective* sees organizational behavior as less predictable and a matter of extremely complex choices (Putnam & Pacanowsky, 1983). They see organizations through the framework of human subjectivity. This interpretivist perspective seems to fit well regarding Internet communication in the workplace. The Internet, computer networks, and media convergence increase the complexity of human communication in the workplace. More choices, more access to information, less privacy, and other Internet characteristics need interpretation to understand. As you proceed in this chapter, we hope you will maintain an interpretive perspective as you try to figure out human subjectivity in the workplace.

Revising Traditional Theories about the Workplace

The field of organizational communication has a rich history of theoretical development. We want to offer ideas about how the Internet changes the basic interpretations of organizations under traditional theoretical understandings.

Bureaucracy

The political and economic aspects of the modern bureaucracy is different from earlier organizations (Weber, 1947). These organizations have fixed rules, stable hierarchy and require specialized training. The Internet flies in the face of existing bureaucracy because the new communication interactions often have no written rules, roles are flexible, and certain individuals may have expertise outside their traditional roles.

Scientific Management

The ideas of scientific management are concerned about the standardization and systems of an organization (Taylor, 1911). This organizational theory is concerned about observing operations so that the organization can effectively carry out tasks. Often criticized for treating humans like machines, advocates of this approach do advocate cooperation and teamwork.

Networks change organizations because the computer system often influences how an organization operates. We know of an organization, for example, which requires the human resources department and payroll department to operate at different times on the computer system network. The organization allocates two weeks for the human resources department to operate, then the next two weeks the payroll department can operate, then the next two weeks the human resources department can operate, and so on. Each of these two departments—responsible for several thousand employees—cannot access computer information half of the time. Are you surprised that recently over 300 people in the organization failed to receive their earned paychecks? Organizational operation—or lack of operation—is determined by that system access.

Administration

The hierarchical approach to administration required workers to view one person in control of the organization (Fayol, 1937). Organization administration requires planning, organizing, commanding, coordinating, and controlling. The work of administrators is less structured than their employees because there is a need to adapt to the work situation, making their workdays varied and unpredictable. Most organizations today are run by people trained in this kind of business administration. You probably can think of examples of innovative e-business leaders, however, who chose a different path to running an organization.

Classical Management

Our modern notions of management come from theories of the self-taught leader who becomes a classical manager (Follett, 1986). We see most chief executive officers as less authoritarian than in earlier decades. The role of managers has become budgeting, coordinating, directing, organizing, planning, reporting, and staffing. These roles are commonly supported today through computer software and network systems.

Human Relations

Theorists became fascinated with human behavior and how changes could affect the way people performed their jobs. Social conditions, participation in management, and human resources departments became crucial elements of this approach to organizations. Interpersonal communication—management who could communicate with their employees—became an essential element in creating effective organizations (Barnard, 1938). We believe that communication on the Internet is primarily a means of interpersonal communication, and subsequently quite influential in organizations. We know an administrator, for example, who uses a daily e-mail to create a sense of camaraderie among employees located in dozens of locations. The administrator's sense of humor and storytelling abilities make the daily e-mails entertaining, while each contains information, directives, or accolades.

Theory X and Theory Y

Theory X believes that people are motivated by material rewards, and basically cannot be trusted and require threats and discipline (McGregor, 1960). Theory Y was a more enlightened approach believing that people were motivated by having meaningful and enjoyable work. Although we have traditionally believed in the value of Theory Y and were quite interested in the development of Theory Z, which has a cultural perspective (Ouchi, 1981), we cannot help think of the fall of Enron, corporate hacking, and similar problems that make us wonder who can be trusted in the Internet age.

Participative Decision Making

Participative decision making showed a concern for people and production (Likert, 1967). The competence and motivation of the employees is as essential as the management of tasks. In this theory, the most effective managers demonstrate supportive relationships, use group decision making, and have high-performance goals and expectations. We see network communications as offering a new way for participative decision making.

Symbiotic Relationship between People and Organizations

The needs of the individual and the needs of an organization must be integrated for success (Argyris, 1990). While the individual and organization need each other, there is a fundamental tension between the two. Individuals must have a sense of self-work and an appreciation of others in the workplace. Modern organizations are often designed to suppress the kinds of interpersonal relationships that people find most satisfying. There is a tension between expressing oneself and being suppressed by the organization. More responsibility, democratic leadership, open communication, and participative decision making can reduce the paradox of the needs of the individual versus the needs of the organization. We believe this theory fits well with our conceptualization of the Internet. The paradox exists, but individuals and organizations can use computer communication to reduce tensions.

Contingency Theories

Organizations need to be prepared to adapt to problems and change. Contingency theories advance a form of systems theory (e.g., Fiedler, 1967; Lawrence & Lorsch, 1967), which works well in viewing human communication on the Internet because they are adaptive. These theories suggest that the following contingencies require an appropriate organizational response:

- Environmental uncertainty, unrest, evasiveness.
- Technological systems.
- Nature and clarity of the task.
- Nature of the leader's power.
- Interpersonal relationships within the organization.
- Conflict or cagey input. (Neher, 1997)

Business Communication on the Internet

Corporate organizations are going through extraordinary changes (Howard & Geist 1995). Perhaps the most powerful influence of the Internet related to the workplace is its ability to act as a change agent. One significant type of change we've discussed, for example, is the idea of globalization. Globalization has changed the nature of many corporations, as they go beyond national boundaries (Barnet & Cavanagh, 1994). But there are many changes happening through business communication on the Internet. If you were to explore through the right databases, for example, you could find thousands of articles and many books about business on the Internet. You can find technology businesses, e-businesses that are accessible primarily through the Internet (also called *dot com* companies), and ways that most American businesses are using the Internet. This book is about human communication on the Internet, and although the business context is an important one, it is just one aspect of human e-interaction. In this chapter we are concerned about the business context.

Remember that a context is the situation, environment, place, or situated meaning. The Internet alters and reduces the significance of the immediate, physical context. In a sense, the Internet transcends the time and place of communication. So how is human communication on the Internet important within the framework of organizations?

Let's start with this definition: An organization is "an ongoing, observable pattern of interactions among people; usually these interactions are planned, sequential, and systematic. . . . Organizations will no doubt become increasingly, rather than less, important to our lives in the future, and so will the roles that people must understand and master in order to be successful in these organizations" (Neher, 1997, p. 14–15).

Systems

A systems perspective of communication theory is one that can be used to analyze computer-mediated communication. The idea with *systems theory* is that entities work together so that the whole entity is greater than the sum of its parts. So it is with the Internet: the Internet operates as a world-wide communication system. The nature of that system affects communication on that system. A systems approach is a relatively open, flexible, holistic, and unbiased way of looking at computer-mediated communication.

Originally introduced by a biologist (VonBertalanffy, 1968), systems theory offers a way of looking at CMC as interdependent computer networks which work together. CMC can work through open systems, in which servers and networks are connected to external servers and computer sources. Or the CMC may be part of a closed system, such as an Intranet system within a particular company. An intranet system is a self-contained system that operates within an organization.

Systems theory values the interrelatedness of parts—subsystems—and the interaction with the environment. We have already made a case for the related ideas of the importance of information and feedback. An open system probably works best (Katz & Kahn, 1978). Using a biological metaphor for understanding a system, one can see how an organization's communication network might be comparable to the blood vessels, with communication being comparable to an organization's blood.

The crucial question is how does the Internet system affect human communication on that system in business contexts? The Internet system controls who can access whom, for example. Some goals of CMC systems are connection, interdependence, speed, access to people, and access to information. These system goals can profoundly affect communication in businesses.

Communication Patterns

When you think of an organization, what kind of communication system exists? Is the organization a *"tall"* organization with a *hierarchy?* A hierarchy works something like this: The employee reports to the supervisor, who reports to the manager, who reports to the director, who reports to the vice president, who reports to the chief executive officer. Maybe you work in a *"flat" organization,* in contrast, were there are few levels of hierarchy and a sense of equity among employees. Communication via the Internet or internal corporate systems leads to a flat organization. The average employee can e-mail the CEO, for example. Information can flow quickly and outside of conventional hierarchical channels. The *communication flow* of messages via CMC can easily go upward (toward the top official) or downward (toward the low-status employees). Telephone hot lines, video and teleconferences, e-mail, computer messages, and similar media can increase the communication flow in an organization.

Computer system communication tends to be brief, even curt. Employees can become overloaded with messages and have difficulty finding crucial information. Employees may not be able to differentiate between essential and nonessential messages. Reading e-mails can create an interruption and invasion of time and privacy. Normal interpersonal cues are absent from communication, and along with that, cues to status and authority are often missing. Access to more information may make points of view less certain and more pliable. As with other paradoxes, we can see values and problems with these various characteristics.

An awareness of the nature of communication via CMC can be useful. When examining how the Internet or a corporate network is used in an organization, you can ask yourself several questions. How rapid and complete is the feedback to messages? Are multiple types of communication modes used? How user friendly is access to information? Is there a Net presence—a social presence—conveyed through mediated communication?

Roles

Communication via networks can confuse the usual *roles* of people in organizations because it changes the way people are able to communicate. Roles are the behavioral expectation in organizations, including position, responsibility, and job. Access to information has long been a source of power in organizations, but access to information via network systems can diffuse that power and control. Internet communication networks are influenced by several factors: Accessibility, centralization, complexity, content, convergence, formality, length of time in operation, openness, power, purpose, and size. Meaning in communication is blurred because roles are blurred by computer-mediated communication. People in organizations function in certain communication roles. Employees can take on any of the following roles:

- Members who usually participate in small group communication.
- Isolates who interact little.

- Liaisons who link people.
- Bridges who connect groups.
- Gatekeepers who control information role.
- Opinion leaders who influence others. (Neher, 1997, pp. 168–169)

There are unique characteristics of communication on the Internet in the business context. Interpersonal communication takes place in a more collaborative way because it is so easy to discuss ideas online, bounce thoughts back and forth via e-mail, and interact with people in different offices. But meaning is blurred because traditional boundaries are blurred too. *Boundaries* are the limits on communication and behavior—the border or edge of what is acceptable—which are expected because of a person's job. In addition, an employee cannot tell when the co-worker is saying something with tongue in cheek, sarcasm, or to be amusing. Without nonverbals, the boundary between serious and fun—the boundary between work and play—is confused. The hierarchical lines of organizational structures are often violated by Internet communication. Fernback (1997) suggests that computer-mediated communication is more than a new way to send and receive messages, the Internet is "a place where people reside in a bodiless form, where social structure and meaning exist, and where action can spring forth" (p. 52).

Communication between people may be seen as a continual balancing act between independence and connection; it is a continual negotiation along such dimensions as power and status; it is a continual construction of borders and social rules enacted through social actions. As such, online communication is a highly charged environment in which boundaries are negotiated and renegotiated as individuals represent themselves within the collectivity. Fernback adds: "The online collectivity does indeed reproduce existing structures, but it also undermines them and raises new possibilities for resistance from the collective and against the culture writ large" (p. 53). While we can apply these ideas to the discussion group online, they also apply to the work group online. Individuals need to find their place within the work group, to both support the group and to be recognized as an individual; work teammates need to rewrite the rules for interaction but may also need to reinforce norms of face-to-face communication (Shockley-Zalabak, 2002).

Internet play may blur boundaries because the employee may play online in the middle of work (Oravec, 2002). Many agencies, businesses, and organizations have created policies designed to curb Internet play during business hours. The employer may decide against regulation, use a permissive policy, or prohibit online play completely. Restrictive policies may create costs for hardware, software, and staff needed to conduct the employee monitoring (Cyber Play, 2000). After reading an e-mail about business, for example, the employee may read an e-mail by a friend, which arrived at the same time. While searching a database for company research, the employee may find an article that relates to a personal interest and spend time reading that article for personal benefit.

This confusion of the work and play boundaries is not necessarily a bad thing and may indeed have positive results. For example, those digressions may provide needed relief and breaks to increase work efficiency or they may be distracting and entice the employee out of the mood to work. The employee may feel guilty for having spent time on personal play while "at work" so may be more inclined to work on business-work while online at home in the evening. Or because the Internet is like play, they may work on the Internet at home because it is entertaining or satisfying. Or the stepped-up expectations about business connection and interaction may prod the employee to work on business every day while away on vacation. The whole, involving process of Internet communication may make work more enjoyable, and the employee may enjoy work "at work" and spend more time doing work while away from the office. There is no boundary between business and personal, no boundary between work and play, no boundary between serious and fun.

You can probably think of examples of how computer-mediated communication can confuse the boundaries of communication patterns and roles. The administrative assistant may have to teach the boss how to work something via computer, making hierarchical ranks inappropriate. The young, newly trained computer person may become a boss to an older, more experienced employee. Communication may take on brief informality in a formal, professional context. Employees may be confused by the changes in traditional communication patterns that are a direct result of computer and Internet-mediated communication. The communication norms, expectations and patterns become complicated and confusing.

Tensions

Paradoxes Abound Regarding the Internet and Organizations

Paradox: After the initial euphoria about Internet companies, the NASDAQ fell. After the NASDAQ fell, online business continued to increase. Paradox: Although the Internet can connect workers all over the country who work out of their home, the workers may feel isolated from each other. In addition to these confusing paradoxes, consider the discussion about the potential for change through the Internet. Paradox: Although we are concerned how the Internet changes human communication, a case can be made for the contention that the Internet changes nothing. The Internet doesn't really change the essence of who we are or change the nature of human interaction. In fact, one can argue that:

- The advances through technology are not that amazing.
- Good interpersonal skills and effective decision making are at the core of human communication, no matter what the role of the Internet.
- The human condition is the focus of communication not the medium of communication (Johnscher, 1999).

Decision Making

Computer-mediated communication is often used to supplement and sometimes to completely replace face-to-face decision making processes of groups in many business contexts. One scholar has suggested four key effects of CMC on business decision making (Neher, 1997), shown in Table 10.1. Not only have business people used e-mail and Internet systems to help communication during decision making, but there is specific software developed to assist in the process.

TABLE 10.1 *Paradoxes of Internet-Based Decision Making*

Fewer restrictions for people in different geographical regions and time zones.	Sequencing can be confusing or lost completely.
Freedom and flexibility.	Lost synergy.
"Leaner" form of communication.	Lacks "richness of the variety of messages... available in FtF situations."
Reduced nonverbal cues "soften status and authority cues, thereby lessening possible inhibitions or constraints on members' contributions."	Loss of leadership may prompt "floundering about, without a clear sense of direction."
Ideas can be analyzed separately from source, so there is more independent thinking and rationality.	Take longer to explain through text.
"Displaying written lists, proposals, vote tallies, and the like on a screen tends to focus discussion more on content than on the source."	Flaming is common because there are no nonverbal softening cues.
More even contribution by all members. "Quiet" members may "talk" more effectively.	Can take longer to solve problems.
Ideas and information more readily available.	System may prompt group to use inappropriate methods.

(Adapted from Neher, 1997, pp. 277–278)

Software can be used to display decision trees, structured group agendas, and similar prepared decision aids. The software package typically includes suggested rules regarding handling of straw votes, the use of decision aids, and the like. (Neher, 1997, pp. 275–276)

Internet Play May Interfere with Business

We have already touched on this concept, but because of intensified interplay, we think the notion of play bears elaboration in the business context. Employees used to gather around the water cooler or coffee pot to chat. In many businesses, the Internet has replaced the water cooler as the social gathering place. While at work, an employee can check e-mail from friends, play an online game, and shop online during odd moments between work. During one business meeting in which everyone was connected to the Internet, for example, one person checked e-mail and another employee shopped online while the boss spoke to the group. In some cases, businesses are clamping down on this kind of employee distraction by using monitoring software and programs that block access to certain types of web sites.

Culture and Climate

The concept of organizational culture comes from systems or contingency theories (Blake & Mouton, 1964). A *culture* is the mores or traditions, which is a persistent, lasting structure, and a pervasive influence on all elements of the organization. You may have heard about the Disney culture, for example, or the Enron culture, which allowed for much corporate success. In contrast, you may have heard about the Enron culture, which created a national disaster. The culture of a workplace includes:

- Cultural network.
- Heroes.
- Rites and rituals.
- Values.
- Work environment. (Deal & Kennedy, 1982)

Culture influences the way people communicate, organizational structure, message behavior, the modes of acceptable communication, and who communicates with whom (see Table 10.2).

Field theory suggests that communication occurs within the framework of a field or background. Concepts of leadership climate and social climate have laid the theory of organizational *climate.* Consider this definition of organizational climate:

> Organizational climate is a relatively enduring quality of the internal environment of an organization that (a) is experienced by its members, (b) influences their behavior, and (c) can be described in terms of values of a particular set of characteristics (or attributes) of the organization. (Tagiuri, 1968, p. 27)

Consider how network monitoring might affect organizational climate. There's nothing new about businesses monitoring employees regarding theft and work performance, but the means for conducting that monitoring takes on new dimensions through electronic means and the Internet. Twenty-five years ago, about 35 percent of U.S. companies monitored their employees electronically, today about 75 percent do so. This monitoring can create a climate of paranoia and fear that "big brother is watching." Some types of monitoring that can be completely hidden to the employee are monitoring devices that can be completed totally by computer—e-mail, computer files, web sites accessed, keystrokes, phone time—and provided in a printed report to the employer. Sales of e-mail monitoring tools for businesses is currently $70 million in the United States, a figure expected to double in the next few years (Asman, 2001). A student in one of our courses could not access her college course online because of monitoring software her employer used to block her Internet access to non-work related activities.

Internet play might be accepted or admonished, depending on the organizational climate. Using business time and resources can have negative impact on businesses that suddenly fail to seem business-like. The Internet may adversely affect what is considered professional behavior in business contexts. An informal e-mail with errors sent to a client may seem friendly or it may seem lazy and unprofessional. The resulting organizational climate may prove quite inappropriate. The construct of climate is that an organization has an ongoing social environment, which influences the way people interact (see Tagiuri, 1968; Lewin, Lippitt, & White, 1939). We often refer to organizational climate as describing the expected behaviors in a corporation, although the concept has been applied to families and other social organizations. We can view Internet communication as a social network, although the Internet is a technology network too. Is there an organizational climate that operates on the Internet? If so, what are the characteristics of that climate? Consider these *factors of organizational climate* that may be affected by human communication on the Internet (adapted from Neher, 1997, p. 93).

> *Structure and Constraint.* Are there formal organizational lines of control? Often the lack of formal lines is a major source of conflict over computer resources.
>
> *Individual Responsibility.* Who in the organization is responsible for CMC? Do individuals follow prescribed uses?
>
> *Warmth and Support.* Do people in the organization use CMC to support each other or as a way to avoid direct interaction.
>
> *Reward and Punishment.* Is CMC used to spy on employees? Are employees encouraged to use the Internet and e-mail system as a personal perk?
>
> *Conflict and Tolerance.* Are problems quickly confronted and solved? Are e-mails used to avoid confronting problems? Is pornographic or other offensive material displayed on computer screens? Are bigoted e-mail jokes circulated that cause a chilly climate?
>
> *Performance Standards and Expectations.* Does CMC enable sloppy work, more informality or imprecision?
>
> *Identity and Group Loyalty.* Does CMC advocate information about who the people are? Do they receive information via CMC that makes them feel positive toward the company? Or does CMC feel like overwork, overload, overwhelmed?
>
> *Risk.* Do people use e-mail simply to cover themselves? Are employers encouraged to take risks via e-mail or do they perceive CMC as a potential source of trouble?

TABLE 10.2 *Paradoxes of Communication Climate on the Internet*

Playful.	Unprofessional.
Friendly.	Inappropriate.
Interdependent.	Resent dependence.
Free.	Confining.
Rules of system.	Violates rules of the organization.
Specific expectations for Internet system.	No penalty of noncompliance with organizational expectations.

E-Business

When we talk about *e-business,* we mean commerce conducted via the Internet, which is also called *e-commerce.* Perhaps you bought this book through an online business such as Amazon.com. If so, you noticed that Amazon.com provides sales from their stock and connects you to private individuals and other vendors. If you walked into a bookstore, you might be able to buy some books that were sold on consignment, but you probably could not begin to access the flexibility offered by an e-business.

No central authority manages the Internet, which enables extraordinary flexibility to e-companies or virtual organizations (Davidow & Malone, 1992). Through computer networking, work groups can easily come and go, or serve ad hoc functions. Further, many companies are outsourcing jobs so that task performance becomes more flexible. Virtual organizations are more likely to experience "adhocracy," or the brief coming together of various groups in the organization (Waterman, 1990).

The Internet can be used to link businesses with their customers to effectively move information and materials. Business can send out targeted e-mails to potential customers and use the Internet to better serve customers, cut costs, and create new revenues. Companies can provide customer services and human resources online, for example, to reduce the number of direct calls they need to handle (Rocks, Pascual, Little, & Brown, 2001).

The Internet can enable corporations to conduct business online. In fact, many e-businesses conduct their enterprise online. Other businesses use the Internet to carry out services. Otis, for example, uses technicians at centers around the globe to monitor their elevators by an Internet link. Otis can monitor and repair hundreds of thousands of their elevators through this Internet-based system.

Literally, the faces of Internet business have changed: Today they are young and old, college-educated and self-educated, small businesses and major corporations, ethnically diverse, and people located around the world. The Internet has intensified the interplay of new people, new markets, new needs, and new products. Small businesses can look like large businesses on the Internet. Individuals can wield power on the Internet. All kinds of people in all kinds of locations have new chances provided by the Internet.

Although the origins of the Internet came from government and education, now the Internet is fueled by corporations in the world economy. Tim, for example, works for an engineering company. The product design and development team is located in Detroit, Brazil, Korea, and Germany. Through technology, Tim's staff meet together once a week—their idea of meeting face-to-face is seeing each other in video images. Their work is transmitted instantaneously through the Internet and computer translations help with language barriers. This kind of interaction is essential in many modern business setups: "The technology gives people an opportunity to collaborate in new product design.... Going forward, we'll give employees more functionality so they can participate in electronic workflows" (Bulkeley, 2000, para. 4, 8). *Electronic workflow* is an example of how the Internet intensifies a worker's ability to interplay with co-workers. In this case, the computer system manages, through software or computer interactions, how people conduct business.

Thus, computer-mediated communication and the Internet can offer new ways to conduct business. Given the cost and concerns about flying, for example, Internet videoconferencing offers a viable alternative. Internet conferencing provides many communication advantages for modern business people. There's no hassle and less expense for the newest online collaboration tools. While still at their desks, colleagues at a distance can talk, share applications, view video of each other, and examine documents in a cross between a conference call, computer connection, and videoconference. At this point, the key problems are voice quality and security, but the collaboration technology will continue to improve (Grimmes, 2001).

E-commerce success seems to have come to companies that blend a physical store with online service. Examples of e-commerce success include: L. L. Bean, Victoria's Secret, Sherwin Williams, and Nordstrom (Blackwell & Stephan, 2001). These retailers have blended the two business forms. There are over 30,000 web shopping sites available to consumers. The decision on where to buy is based on the site quality, trust, and positive feelings toward the site. Online web sales should be about $78 billion in 2003, and this requires that business people figure out global strategies for their marketing. More than 20 percent of U.S. citizens have bought online, primarily from U.S. businesses. Online sales success seems to be based on the ability to guide traffic to the site, obtaining an initial sale, and enabling repeat business through a positive experience. Some dot com companies have spent up to 50 percent of their total budgets on advertising. One extreme example was a company that spent millions of dollars on banner advertising, which yielded $100,000 in sales. The average Internet e-business spent $35–$150 on marketing and promotion per first time customer, which makes the return of that customer essential (Lynch, Kent, & Srinivasan, 2001). These statistics suggest that there are still many ups and downs as e-businesses try to figure out how to conduct business via the Internet.

Online Business Interaction

This book cannot begin to raise every issue about business communication on the Internet, but certain research findings may help prompt your investigation and subsequent understanding of the topic. This section presents some research findings which relate to human communication on the Internet. The objective is not to present every idea about human communication on the Internet in a business context, but to provide an array of ideas that may prompt your own investigation and understanding.

Consumer Communication

Online businesses collect extensive data in an effort to analyze who you are, what you want, and what will make you buy. Your use of the Internet communicates information about you to many businesses.

> Academics and practitioners alike have been arguing about whether the Internet brings a revolutionary change in the fundamental way we do business or if it simply offers a new distribution channel and communication medium. Regardless of the answer to that debate, one thing is for sure: the Internet provides managers with an enormous amount of customer

information that was previously unavailable. Thus, the new struggle has been to manage this information and to accurately and efficiently use it to somehow measure customers, trends, and performance. (Moe & Fader, 2001, para 1)

Interpersonal Communication

Interpersonal communication—one to one communication—has a place in business contexts. Consider for a moment, how computers have shaped interpersonal communication in the workplace. We know that the quality of work completed through computer-mediated communication can be as good or better than when people work face-to-face (Reid, Malinek, Stott, & Evans, 1996). We know that in today's workplace, people tend to communicate through diverse means that include emerging technology (Haythornthwaite, Wellman, & Mantei, 1995). We know that online communication can be used to replace and enhance other forms of communication. There are interesting paradoxes within the ideas of scholars investigating e-mail communication (Dickinson, 2003). And people using e-mail in the work contexts need to be particularly careful regarding the clarity, appropriateness, and security of messages. The businessperson will want to understand, for example, the nature of communication in computer conferencing (Angeli, Valanides, & Bonk, 2003). One study, for example, found that the face-to-face communication in an organization decreases as e-mail use increases. The kind of communication lost was the more casual and playful social communication that does much to create a positive climate in an organization (Sarbaugh-Thompson & Feldman, 1998). There are at least two likely explanations. First, because of the ease and flexibility of online communication, people send e-mails instead of walking down the hall to give someone information. So, because they stay in their office they have fewer casual conversations. Second, because of the playful characteristics of some Internet communication, employees may be doing as much playful social communication, but it's in the form of fun e-mails. And with the blurring of boundaries, is there a blurring between personal and professional relationships?

Collaborative Communication

We have discussed the concept of Internet collaboration elsewhere, but the concept warrants discussion in this chapter. Internet technology has increased business' ability to have more people involved in decision making and team cooperation that spans the miles. That collaboration can also create problems.

One advantage of online collaboration is the reshaping of time and space. A company can have offices around the world, and through e-mail, employees can communicate as though they are in the present. In fact, there is no traditional time or space in this mode of communication (Sproull & Kiesler, 1991) because time and space are altered in social and psychological ways (Kasket, 2003; Stix, 2002). In fact, when people communicate via the Internet, they build their own concepts of time and space in a physical sense and in their minds (Jamieson, Fisher, Gilding, Taylor, & Trevitt, 2000; Teske, 2002).

Internet and hardware and software projects can be extremely complex and expensive. They require integrating the ideas of an array of staff, are cost intensive, can take more time than what is desirable, and the complexity can lead to mistakes. The problem that develops in this case is somewhat like groupthink (Janis, 1982).

Groupthink

Groupthink is a problem that develops when a cohesive group makes bad decisions; when getting along and going along with others is more important to the group members than debate. In an effort for collaboration in their work together, the group may begin to take an unrealistic view of the problem. As they invest in a project, they cannot back out without negating the importance of the group, the group's cohesiveness, and the planned project. Particularly important in the case of Internet and computer communication is the huge investment in effort, time, and money in IT projects. That investment makes the members even less likely to stop the momentum of a poor project (Smith, Keil, & Depledge, 2001).

An interesting finding is that through Internet communication, the individual may seem less important than the team. In computer-mediated groups, participants are more apt to confuse who said what than in face-to-face groups (Durso, Hackworth, Barile, Dougherty, Ohrt, 1998). This finding is understandable because of the limited source of information and the fact that participants may simply read text without trying to attach who is the source of the information. Yet when an e-mail comes from the boss, they may sit up and pay attention.

Although the ideas of team, workgroup, and collaborators are not exactly the same, the idea of using multiple employees to work together to create and accomplish tasks is popular in today's businesses. The Internet provides the advantage of making people separated by distances, locations, and telecommuting employees more accessible to work groups. A *partially distributed work group* is one with a core of employees linked to other members through the Internet. In general, these groups function much like groups which are all together physically. The group member's attitudes toward the mediated communication of the group improves over time (Burke, Aytes, Chidambaram, & Johnson, 1999). While collaborative work online can be engaging and successful, online work can flatten the typical organizational hierarchy, and lead to social shifts in an organization (Ashling, 2003; Henry, 2002).

Communication Overload

One of the serious problems with business communication is the sheer volume of information exchanged. The number of e-mails a business executive receives and the availability of relevant information for decision making can be overwhelming. The problem with overload is that employees may not be able to distinguish what communications are worth reading and what should be ignored. Consider, for example, spam, which can fill an e-mail in-box, making it hard to find important messages. Spam is unsolicited e-mail sent to a number of people, rather like junk mail in your postal mailbox. Spamming is a real privacy issue for Internet users, who may resent the flood of unsolicited information.

> Survey.com polled 1200 Internet users for the Coalition Against Unsolicited Commercial E-Mail. These users receive an average 24.11 e-mail messages per day, 39 percent of which they considered to be spam. Fifty-one percent said they appreciated e-mail from companies with whom they have a relationship; and 25 percent were neutral. Seventy percent said they disliked getting mail from companies they did not do business with. They had the following negative reactions: 35 percent opened new e-mail accounts; 40 percent refused to con-

duct business with these companies; and 60 percent felt that junk e-mail was not useful, while 3 percent thought that it was. (Stead & Gilbert, 2001, para. 15)

Risky Communication

There is always the potential for risk in communication. When you tell a coworker about your inner feelings, there is the risk that the individual will think you have crossed a line in your work relationship. When you tell your boss that you think her latest idea has a serious problem, there is the risk that she will feel defensive and react negatively to you. Communication through computers creates a new type of risk in business contexts.

> These new threats are the direct result of our increasing dependence on computers. Most of the U.S. economic gains in the last decade were made possible because of the increased productivity resulting from technology. The most obvious of these developments—the Internet—provides the unprecedented power to reach millions of people instantaneously, forever changing the face of business.
>
> When employees and customers depend on uninterrupted access to mission-critical data, the consequences of downtime are magnified. So is the cost. Today, one hour of network downtime can cost, as one CFO puts it, "Eight million dollars of commerce that didn't happen." (Simmons, 2001, para. 2–3)

Skill Section: Manager Communication

Kent (2001) offered an interesting metaphor for the Internet by suggesting that because of the consumerism elements of the Internet, a managerial metaphor should replace the current spatial/relational metaphor. Perhaps a managerial metaphor is useful. Does the Internet shape the way management interacts with employees? Or does management determine the way employees interact because of the Internet? Both. Consider the findings of one study, which suggest that many managers use e-mail negatively, to exert their authority and pressure employees (Lamude & Larsen, 1998). Another study suggested that people communicate better regarding tasks depending on their experience level, so that inexperienced computer users performed better face-to-face, while experienced users performed better in computer-mediated communication (Adrianson & Hjelmquist, 1999). No surprise there. Also, the Internet can be the connection when a manager has to be away from the main office.

> Occasionally you find yourself managing from a distance. These types of situations can present an array of challenges from maintaining communication with your staff to ensuring your team has the resources needed to complete projects in your absence. Developing a strategy is the key to remaining effective as a manager while you are out of the office. Some tips include:
>
> 1. Prepare for the unexpected.
> 2. Stay connected.
> 3. Empower your employees.
> 4. Reassess the situation. (Messmer, 2001, para. 1)

TABLE 10.3 *Paradoxes of Business Communication Climate in Internet Contexts*

Meaning blurred.	Emphasis on meaning of words.
Boundaries blurred so business roles are unclear.	The public and individuals in an organization have easier access to each other through communication on the Internet.
Communication patterns are changed and confused by the Internet system.	People may use the Internet to circumvent established lines of communication.
Internet changes everything, including the nature of business.	Internet changes nothing about effective business communication.
Internet provides individualized opportunities for young and new entrepreneurs.	Internet businesses have failed.

Reflection

What paradoxes affect human communication on the Internet in business context? (See Table 10.3.)

How will you use the Internet in your professional life? Will your meaning be clear or misinterpreted? Will your personality and spirit come through in what you say online to coworkers?

Corporate America is an amorphous entity today, defined more by media and information flow than physical reality. The previous barriers of gender, age, ethnicity, and geographic location dissolve in formats of online collaboration.

> I believe that our evolution toward an ever-richer information context will forever change the way that we define ourselves, our class, and our view of others. I also believe that this information richness will create greater precision in class definition and will obviate the cumbersome affiliations based on race, gender, or national identity. I fervently hope that the greater availability of information will end coalitions formed on the basis of demagoguery; although the demagogue has equal access to the Internet, I expect that the pervasive availability of diverse ideas and interpretations will allow people to form better affiliative choices. (Townsend, 2000, Reflection, para. 3)

Reflect on the value of equal access for corporate America.

Case for Discussion

In your course discussion group, give your analysis and perceptions on the following business information.

One business saved $60,000 annually in travel expenses by using Internet conferencing instead of face-to-face meetings. Such savings can make a significant impact for small businesses. For larger companies, Internet profits are even greater. DuPont, for example, saved $400 million a year by switching to purchasing their supplies online. DuPont's purchasing system was streamlined for all employees through a web site login that allows employees to buy whatever they need (Rocks, Pascual, Little, & Brown, 2001, para. 6, 13).

In your course discussion group, research and discuss the concept of online shopping.

A wide variety of people in the United States buy online. The highest concentration of online buyers are 50+. Buyers who are 65+ are the fastest growing group of e-buyers (Burwell, 2001). Surprised? E-commerce has increased 30% since the 2001 Nasdaq slump began (Golson, 2001, p. 14). These findings are inconsistent with what most people think about Internet shopping.

A paradox is a set of ideas or principles which are contradictory. Yet often in CMC, contradictory behaviors and expectations co-exist. Paradoxes can even merge or exist interdependently. Discuss merging Internet paradoxes in organizational communication. In 2005, 941 million people will use the Internet around the world ("By the numbers," 2002).

Internet Investigation

1. *Latest Statistics.* In 1998, more than 75 million U.S. Americans were online (Suler). Find out the current statistics. How do you explain the rate of growth of online use? What is the most influential business communication on the Internet? How do you think the Internet affects communication in business contexts? Write an essay on this subject, which you can post to your online discussion group.
2. *Business Web Sites.* Find web sites for examples of very small, medium, and large corporations. What do the businesses communicate through their web sites? Can you think of ways the businesses are communicating more effectively through the Internet than they may have communicated prior to the Internet?
3. *Cultural Influences.* U.S. workers spend more time working than people of any other nation. We now surpass the Japanese in hours worked per year. The computer, which was supposed to reduce workload, has been blamed for increasing work expectations. Investigate how the computer affects the way people work in different countries and post your findings to your online discussion group.
4. *Media Business on the Internet.* What is the nature of the business of mass media on the Web? Investigate television, newspapers, radio, and advertising businesses on the Internet? How is Internet media different from conventional media? Who is attracted to reading the newspaper online, for example, instead of reading a hardcopy of the newspaper? How does streaming audio or video compare to conventional broadcasting?
5. *E-mail Investigation.* You probably are not surprised that in business contexts, some people prefer the speed and flexible time of communication through e-mail in preference to other modes of communication (Murray, 1991). Design a questionnaire that will gather information about how people use e-mail in business contexts. Look online to see what information you can find on the topic. Investigate the advantages

and problems attached to e-mail use in businesses. You may want to work in pairs or small groups to develop your measure. Conduct a survey of six business people and share the information with other members in your class.

6. *Whistle-Blowing.* Investigate some of the policies about whistle-blowing. You may find different approaches in the United States and the United Kingdom, for example. If you observed illegal activity in your organization, what would you do?

7. *Diversity in the Multicultural Workplace.* One element of the changing nature of the workplace is the increased presence of women and minorities. People who are able to communicate with people of different cultures can increase their effectiveness in the workplace. Effective intercultural communication has many potential benefits, including healthier communities, improved commerce, reduced conflict, and personal growth through tolerance. Cultural sensitivity is a necessary skill for success today. Investigate the current trends in the workplace, which reflect the changing nature of employees. How can you best adapt to those changes?

8. *Romance in the Workplace.* Most communication experts advise students to avoid romance in the workplace. One problem is if the relationship fails as a romance, the two people still have to make the work relationship work. Other potential problems include power, sexual harassment, and confidentiality difficulties. Yet Lenny met his wife where he worked, and Joan works for the same employer that her husband does. You probably know people who maintain a successful romantic relationship while working together. Investigate online web site and database articles on the subject of romance in the workplace. Write a list of rules you can live by. You may benefit from doing this assignment with other members of your class.

9. *Ethics.* Find examples of ethical violations in corporate America. The Enron case is among the most devastating. How have computer systems been involved in those situations? What can be done to protect employees, management, stockholders, and the public from unethical corporate practices?

Concepts for Analysis

We have provided concepts relevant to this chapter so that you can check your understanding. These concepts are worth dissecting in more detail. As a research option, examine one concept. Use our citations as a starting point for your study. We recommend that you search for refereed articles in an online database. Find at least three solid references you can read. Summarize, scrutinize, and present your analysis to other people in your course.

Questions to stimulate your inquiry: Do you agree with the ideas as we presented them? What other points of view should be considered? What details need to be added to fully understand the concept? What can you contribute through your personal analysis of the concept? What more do scholars need to know about the topic?

The *hierarchical approach* to administration of organizations looks at hierarchies, where workers view one person in control of the organization (Fayol, 1937).

Boundaries are the limits on communication and behavior—the border or edge of what is acceptable—which are expected because of a person's job.

A *bureaucracy* is a political and economic structure frequently associated with parts in a system that fail to communicate effectively with other parts, creating unnecessary encumbrance (see Weber, 1947).

Business context is the situation, environment, place, or situated meaning of business or professional communication.

Classical Management notions come from theories of the self-taught leader who becomes a classical manager (Follett, 1986).

Climate in an organization "is a relatively enduring quality of the internal environment of an organization that (a) is experienced by its members, (b) influences their behavior, and (c) can be described in terms of values of a particular set of characteristics (or attributes) of the organization" (Tagiuri, 1968, p. 27). *Climate* is the ongoing social environment of an organization, which influences the way people interact (see Tagiuri, 1968; Lewin, Lippitt & White, 1939).

Communication flow of messages can go upward (toward the top official) or downward (toward the low status employees).

Communication overload is a serious problem with business communication and is when the sheer volume of information exchanged overwhelms the employee. A large volume of e-mail, for example, can cause information overload.

Contingency Theories suggest that organizations need to be prepared to adapt to problems and change (e.g., Fiedler, 1967; Lawrence & Lorsch, 1967; Weick, 1995).

The *critical perspective* in studying organizations is primarily concerned with power and control in organizations. The critic is more concerned about why the organization is running effectively. Who benefits? They may be concerned with material or humanist outcomes.

Culture in an organization is the mores or traditions, which is a persistent, lasting structure, and a pervasive influence on all elements of the organization.

Dot com companies are Internet businesses.

E-business is commerce conducted via the Internet, also called e-commerce.

Electronic workflow is when a computer system manages—through software or computer interactions—how people conduct business.

Flat organization is where there are few levels of hierarchy and a sense of equity among employees.

The *functional perspective* examines the effectiveness of organizations. It is concerned with how organizations operate efficiently and productively.

Groupthink is a problem that develops when a cohesive group makes bad decisions. In an effort for collaboration in their work together, the group may begin to take an unrealistic view of the problem. As they invest in technology resources, for example, they cannot back down because of the huge investment they have made.

A *hierarchy* is a "tall" organization, meaning it has a clear chain of command.

Human Relations approaches to organizations are fascinated with human behavior and how changes might affect the way people performed their jobs. Social conditions, participation in management, interpersonal communication, and human resources departments are crucial elements of this approach to organizations.

Interpretive perspective sees organizational behavior as less predictable and a matter of extremely complex choices (Putnam & Pacanowsky, 1983).

Partially distributed work groups are ones where a core of employees are linked to other work group members through the Internet.

Participative decision making shows a concern for people and production (Likert, 1967).

Professionalism is when an individual uses the Internet in a business-like way, using etiquette and appropriate formality and accuracy.

Roles are the behavioral expectations in organizations, including position, responsibility, and job.

Scientific management ideas are concerned about the standardization and systems of an organization (Taylor, 1911).

A *symbolic relationship* balances the needs of the individual and the needs of an organization as they become intergrated for success (Argyris, 1990).

Systems theory suggests that entities work together so that the whole entity is greater than the sum of its parts. The Internet is more than each computer on the system, the Internet is a system.

Theory X states that people are motivated by material rewards, and basically cannot be trusted and require threats and discipline (McGregor, 1960). Theory Y is a more enlightened approach believing that people are motivated by having meaningful and enjoyable work. Theory Z takes a cultural perspective (Ouchi, 1981).

11

Educational Contexts

Every few years there seems to be some new push in education. In the 1990s, faculty and students had to find new ways to assess learning and performance because politicians thought the array of tests previously used were not good enough. Americans tried *outcomes-based education,* or a results orientation (Hill, Lake, Celio, Campbell, Herdman, & Bulkley, 2001; California Department of Education, 2002). Schools tried viewing the student as consumer and giving them what they want, regardless of what faculty believed constituted appropriate learning. And teachers were being held accountable for what they did in the classroom, as if they weren't accountable in the last week, last year, or last century. These mandates are prompted by the private sector, often through legislative actions and usually out of frustration because education is failing their constituents in some way. The mandates appear less about the schools' failures, however, and more about the U.S. public's increasing lack of trust in public institutions (Penn, 2000). Although, it must be noted that an accountability movement is evident in Canada as well (Education Act, 1990). Because much of the current controversy about education reform is motivated by technology, we must analyze the complicated and challenging issues of the Internet (Cardenas, 2001). An educational context we want to consider is how the Internet shapes learning. In this chapter, we will consider how the Internet actually can intensify learning. One of the objectives for modern education through the Internet is to permit anyone to learn anything, anytime, anywhere. The Internet is full of paradoxes, including ones that involve intensification of effort, controversies in policy formulation, and other educational disagreements.

Focus

```
TO: Students
FROM: paradoxdoc@hotmail.com
RE: Overview: Education
```

> Lenny and Joan contend that the Internet can intensify learning. They explore the role of the Internet in distance education, where faculty and students may communicate via the Internet without ever actually meeting face-to-face. The Internet allows anyone to learn anything, anytime, anywhere. The Internet results in an intensification of educational effort.
>
> Lenny and Joan ask you to think about educational policy and Internet communication, and the controversies surrounding those policies. They discuss how the Internet is used for academic collaboration. There is research that shows positive effects of Internet communication in educational contexts. New books, courses, and popular magazine articles have attempted to improve computer and Internet literacy. For instructional purposes, students and teachers have employed the Internet to benefit by: (a) aiding student learning, (b) presenting material, helping the learner master concepts through drill and practice, (d) preparing the learner to perfect skills through problem application, (e) providing independent, individualized, flexible, and accessible instruction, (f) permitting classroom and online demonstrations and models, and (g) accessing, monitoring, recording, and managing information. For students, Internet-mediated communication seems to have many positive effects. What do you think?

The focus of this chapter is on educational contexts, specifically how the Internet has created paradoxes about education. We are particularly interested in higher education although our comments are not limited to higher education. At the end of this chapter, you should be able to answer the following questions.

1. Is learning intensified through the Internet?
2. Can the Internet be used to allow anyone to learn anything, anytime, anywhere?
3. How does the Internet intensify effort?
4. What policy controversies are prompted by the Internet?
5. How does the Internet enhance collaborative communication in education?
6. What are the paradoxes of Internet communication in education?

Intensified Learning

Is learning intensified through the Internet? The use of the Internet in higher education operates on a continuum ranging from classroom support to distance education. Faculty can provide online *supportive education* through web pages, e-mail, and mailing lists (Hanson & Jubeck, 1999). You can think of this level of online instruction as supplementing the course in ways similar to using handouts, faculty-student study sessions, individual appointments, and readings.

In *distance education,* faculty and students may communicate via the Internet without ever actually meeting face-to-face. Students may take such courses from home and teachers may operate from their home or office. Distance education is when students take courses away from the campus through mediated communication, such as videos, television, and the Internet. How can technology be used in distance education? Messages may be sent and received live (i.e., with people at their computers during the same time period—synchronous) or at different times (asynchronous), like leaving a message on someone's telephone answering machine and getting an answer back on a telephone answering machine.

Not all distance education is Internet-mediated communication, because the distance education course can be conducted through regular mail, televised courses, and other methods. Not all Internet communication courses are used for distance education because CMC is most often used to assist conventional course instruction. In both supported and distance education classes, teachers and students may operate within an online course environment (e.g., a kind of electronic mall containing course documents, online syllabus, electronic groups, discussion boards, chat rooms, e-mail, a gradebook, online testing, automatic grading, homework drop boxes, and slide presentations). In short, *online education* refers to both computer support and computer distance education.

While online instruction varies from supplementing a classroom course to instruction one-hundred percent at a distance, these are very different experiences in many ways. The point here is to recognize that the popular term, "online," can apply in both instances, supplemental and distance education. It is a matter of degrees. We have taught courses both ways, as well as the hybrid form of instructional television plus online. Each has its own characteristics, and each affects communication. We will focus on the general question of whether or not online education, in any of its forms, adds value to education.

Students seem to learn effectively via online and conventional techniques, but do the advantages of online flexibility outweigh the disadvantages of reducing face-to-face contact? While some educators see technology as a miracle cure for problems, other faculty want to avoid technology completely (Delacote, 1995), and there are faculty whose attitudes reside between those extremes. One disadvantage that often bothers faculty is that online instruction lacks the incessant nonverbal communication of the classroom. "Even the student who's totally clueless sends out an SOS with every glassy-eyed blink" (Kilian, 1997). In contrast, a teacher may not know when an online student is having trouble.

Research has typically found no significant difference in learning outcomes for students in the online, distance education versus traditional classroom settings (Phipps & Merisotis, 1999; Wegner, Holloway, & Garton, 1999). At the same time, much of the research on this issue has flaws (Gold & Maitland, 1999) because perhaps the research is based on academics using technology and distance learning who are creative faculty able to teach well via any means. An additional weakness in much of the research has to do with selection of students (subjects) into classes. With self selection of students, we are likely to find non-random placement of students into comparison groups. Regardless, an analysis of 355 research studies found no significant difference between face-to-face learning and technology-based distance learning (Russell, 2000). Of course, you could point to specific studies on both sides, those that find higher scores for students in traditional classroom courses and those that find higher scores for students in online courses, on a range of measures (The "significant difference phenomenon," http://teleeducation.nb.ca/significantdifference/). Some research suggests that a combination of the traditional face-to-face classroom and online learning produces a better outcome than face-to-face alone (Chadwick, 1999; Chadwick & Callaway Russo, 2002).

Allow Anyone to Learn Anything, Anytime, Anywhere

What are your first thoughts about online education compared to traditional classroom education? Have you ever taken a course that was supplemented by a web page or electronic discussion group or the Internet? Have you ever taken an online distance education course? What would be some advantages or disadvantages you would expect to experience in taking an online, supplemented course? What would be some advantages or disadvantages you would expect to experience in taking an online, distance education course? More and more we hear about courses offered on the Internet. The number of courses offered this way is growing. Increasingly, in online courses, the class never meets face-to-face. Some people think this instruction is the best thing that has happened to education in a long time and others think it is a bad idea. What are your thoughts? When students enrolled in web-supported courses were asked what they thought about online education versus traditional classroom education, their responses were varied. Here is what one student had to say.

> I feel that there is certainly potential for positive contributions, but I find myself being skeptical at the same time.

This student expressed ambivalence about online education. Another student thought that online education was a very good thing, and he explained (including some mistaken ideas about what can and cannot be done online) why:

> I believe that it is much easier for me to take courses over the Internet. The class atmosphere is totally different because you can do it from the confines of your own home. No longer will we have oral presentations or group projects, yet who will miss those dreaded oral presentations which cause major stress on yourself. And those classroom tests? God how I hate taking tests in the classroom, where it seems the professor is lurking over your every twitch. With technology expanding the way it is today, I don't think that we should have any trouble adapting to courses offered through the Internet. I personally feel com-

fortable sitting behind the monitor and accessing anything I need to without a problem. I do realize that not everyone feels comfortable using computers or the Internet, but it is time for those people to get connected because there is no way that we are ever going to go back to snail mail or anything like that. The information age is upon us. Now you must choose, are you in, or are you out?

One feature of online education that stood out as an advantage to all students was the convenience of taking a course (or part of a course) from home, although many students were worried about the loss of social learning by not being in the classroom. Students also liked being able to work on a course when they wanted to work (Shedletsky & Aitken, 2002).

While some students preferred one way or the other, traditional classroom versus online distance education, many took a position in the middle. The fence-sitters saw good and bad in online distance education. Interestingly, although people call online education both a good thing/bad thing, the growth of online education is exploding. The U.S. Department of Education's National Center for Education Statistics reported that the use of Internet-based technologies in teaching grew 22 percent between 1997 and 1998 (Lewis, Snow, Farris, Levin, & Greene, 1999). For the same time period, the survey also pointed to a 33 percent rate of growth among higher educational institutions offering courses through distance education. With each passing year, the rates increase. Cattagni and Farris (2001) reported that "98 percent of public schools were connected to the Internet (compared to 35% in 1994) and 77 percent of instructional rooms were connected to the Internet (up from 63% in 1999). In 1994, only 3 percent of instructional rooms were connected." Being connected to the Internet is not the same thing as online education, but when students interplay with resources and people online, the Internet influences classroom learning. The Master Plan for Education in California, released in the summer of 2002, makes it clear that access to the Internet is a significant goal of the plan (Joint Committee to Develop a Master Plan for Education—Kindergarten through University).

The idea of "anytime, anywhere" learning technology is a vision taken seriously by the public schools, K–12. For instance, a task force on technology and the future of teaching and learning in the state of Maine took a strong stand in favor of integrating the Internet into public education (Teaching and Learning for Tomorrow: A Learning Technology Plan for Maine's Future):

> If technology is a challenge for our educational system, it is also part of the solution. To move all students to high levels of learning and technological literacy, all students will need access to technology when and where it can be most effectively incorporated into learning. With the guidance of good teachers with technological facility, computer technology and the Internet can provide students with a pipeline to explore real world concepts, interact with real world experts, and analyze and solve real world problems. Computers and the Internet offer the potential to keep classroom resources and materials current with the contemporary world to an extent that is unprecedented. Computer technology also offers opportunities for self-directed, personalized learning projects that can tailor the curriculum to student interests and engagement, and allow teachers to facilitate active student learning rather than merely the rote transfer of information.

Although some educators question the value of Internet resources, their schools are racing to buy the equipment to make online education possible. A survey of the use of technology in U.S. colleges and universities showed a significant trend toward spending on distance learning ("Year 2000," 2000). Internet access is spreading on college campuses. Workshops are offered. Support staff are the prima donnas of information. Some teachers are caught up in investing work in online projects. Universities boast to their first-year students about their smart classrooms (high tech) and large number of Internet-based courses (support or distance education).

> The overall societal expansion of CMC is dwarfed by the growth of CMC on college and university campuses. The use of listservs, web pages, and other CMC modalities is now common place in the academy. Computers are now the rule rather than the exception (McCollum, 1998b, p. A27). Many colleges and universities are now eagerly working to prepare experts who can facilitate use of the Web and extend employment horizons for students (Kunde, 1998). College instructors are also being requested to extend their use of advanced computer technology, including "essay assessors" to help score student writing (Perlstein, 1998). Those in higher education continually exert efforts to promote CMC and often promote the Internet as having unique supplementary assistance for classroom instruction. All such efforts within higher education has a "trickle down" effect on the general population. (Lane & Shelton, 2001, p. 243)

In K–12 teaching, we see a trend toward infusing technology into *preservice teacher education* (Abdal-Haqq, 1995). The Maine Department of Education, for instance, announced that "Internet skills are the new '4th' basic in education" (Maine Department of Education). Again and again, however, as often as we find examples of treating online pedagogy as worthy of extra expense, training, time, and effort, we also find examples of perceiving online pedagogy as a threat to education. One report summed up this paradox: "In reviewing the literature on distance learning, one quickly discovers both hyperbole and deep skepticism" (Schmidt, 1999). We can point to prestigious universities that have adopted online pedagogy: Stanford offers a full engineering degree, and Duke offers a full MBA online (with some face-to-face meetings). Fully online classes now exist at Oxford and Harvard. But in counterbalance again

> We hear of concerns among faculty about job security and the implications for promotion and tenure as well as reward structures, concerns about the content quality of distance learning, and a series of technical issues such as intellectual copyrights, accreditation, transferability of credits across institutions, and the integrity of undergraduate and graduate programs of study. (Schmidt, Shelley, Van Wart, Clayton, & Schleck, 2000)

One pattern has been consistent in the history of communication technology: We have been wrong about how people would use each new medium of communication (Head, 1976). Not until we actually set people loose on the Internet do we discover just how they will behave. While we may design systems and build plans for how to profit from a new medium, make it available, anticipate its advantages and disadvantages, we find we need to redesign after we see how people use the new medium. And why should we have thought otherwise? Anyone who has ever carried out empirical research knows the importance of a pilot study to learn of the unexpected responses—the ambiguities that hadn't surfaced in instructions or questions.

Those of us who have created Web-based courses, who have used e-mail in our courses, who have used electronic discussion groups, or have had students drop their homework off electronically, have some idea of what is involved and what forces affect these endeavors. Those who have engaged in online work know something about the (great) effort involved, the (positive) student response, the (mixed) reactions of colleagues and administrators, and even the often difficult give-and-take of interacting with the technical-support staff. They know the planning and work it takes just to reserve the computer classroom. Educators who have engaged in online work have experienced the contradictions and politics of that work. They have witnessed both praise for using technology and criticism of their online work, the encouragement to do online work and the failure of support.

The online researcher revels in the convenience and efficiency of online literature searches, but suffers from information overload. Technology users can swing the door wide open to the world for their classes, only to have it slammed shut by computer crashes beyond their control, errors in scheduling computer labs, or failure to find someone with a key to open the door to the computer classroom. Or, faculty may prepare to teach a new course that requires particular software, only to be told that the institution has decided not to purchase that software. And, of course, there is always the threat of a power outage. The larger point is this: Technology offers powerful extensions of the ability to acquire and disseminate information, but it also constrains those extensions and intensifies decision making errors.

Intensification of Effort

How does the Internet intensify effort? Originally, computers were touted as work savers. In fact, they have raised expectations about the amount and quality of work expected (Breen, Lindsay, Jenkins, & Smith, 2001). Proposing to teach a course online invites the reaction that the course can now seat more students than the traditional classroom. To the casual observer, online teaching appears to resemble mass communication, wherein one person can speak to many. The difference, of course, is that with online teaching, the many speak too, and they usually speak much more frequently and longer than in the conventional classroom. Online instruction is essentially a writing or text medium, resulting in a writing-intensive course. Teaching online also places more pressure on the teacher to respond than does traditional teaching (Gaud, 1999). The Internet enlarges the classroom because the teacher is accessible all the time. Thus, there is an increase in the need to read e-mail from students and respond to them. Faculty really need fewer students in their courses when teaching online because there is so much reading and writing. Students responding to discussion questions on the Web require greater teacher involvement than in the traditional classroom: "Unless they get direct feedback to their postings—[students] do not know whether the instructor is reviewing their comments and evaluations" (Miller & Jones, 1999, p. 45).

On the down side, we hear about the intensity of the work for online students and faculty.

> Often, I would find myself drowning in text overload. Even with only nine students, sometimes I'd find 50–100 messages in the backlog if I had taken off a few days from the conferencing system. I found that if I didn't make a habit of logging in at least once a day, I would fall behind in reading and processing messages. Since I had so many other things competing for my time, I'd often find myself frantically logging on the day of a deadline to finish an assignment. (Gale, 2000, p. 53)

The same pressures exist for faculty and students. Although an enthusiastic teacher, one faculty member expressed dismay when he found more than 200 student e-messages one day. Kalmbach (1997) explained the workload problem as keeping up with the relentless pace of change, more powerful computers every year, and new versions of software (p. 101).

Both faculty and students find that they need to participate in online courses by spending time reading and writing to the discussion site. Both faculty and students find the experience can be quite labor intensive. The pressures prompt some teachers to establish guidelines specifically for online interaction, which expect students to articulate their thinking processes and value multiple perspectives while resolving conflicts. Students also may be required to state their intentions "using a mix of probes and supportive comments to extend conversation beyond simple question-answer interaction, and using conferencing software effectively" (Sherry, Billig, & Tavalin, 2000).

Nonetheless, many people outside of the online world mistakenly think that the online course is less demanding than the face-to-face course. Teaching online requires teacher training, even if it is self-training. Consider this observation about Internet learning: "While technology evolves quickly . . . the human ability to understand, shape, and incorporate these changes evolves slowly. . . . The time required for the development, refinement, and adoption of appropriate new pedagogies may be a decade or more" (Feldman, Konold, & Coulter, 1999). Teachers need to change instructional patterns, prepare new materials, and invent new ways of being effective (Lieb, 1999). At the same time, teachers are asked to prove that online learning is at least equal to or better than classroom learning. Classroom learning is the *norm*. The implication is that online learning is the *abnormal*.

While teachers are learning how to teach effectively through the Internet, the technology is undergoing rapid change. New software is being developed, new sites are coming into existence, new uses are popping up all around, new hardware is being invented—teachers feel they must sprint to keep up:

> One of the most difficult challenges for anyone teaching in a computer-supported classroom is to keep up with the relentless pace of this change. Computers keep doubling (and even tripling) in power every year, and publishers keep producing new and purportedly better versions of their software. As soon as we become comfortable with a new architecture or a new word processing program, something newer, more exciting, and potentially even more useful appears. (Kalmbach, 1997, p. 101)

All of this change amounts to more work.

Questions also arise as to how many students constitute a section; whether courses have to be offered in semester slots; and how teachers will be paid for preparation time, since preparing materials for online courses is labor intensive. Consider how these scholars grappled with educational controversies.

> "A lot of them are doing it because it's new and different and fun . . . . The question is for how long. A lot of online distance learning is being done on the good will of faculty. But that's not a solid foundation." (Weiner, 2000)
> "The maximum enrollment in a single Web class section should be no more than 32 students with grading support, and 16 without graders." (Gaud, 1999)
> "Although the innovations are usually perceived as technological, the actual innovations are technology-enabled pedagogical approaches such as experiential learning in collaborative learning communities. Technology is only the enabler, and no single technology is sufficient on its own." (Cox, 1997).

Educational Policy and Internet Communication

What policy controversies are prompted by the Internet? We won't pretend to be insiders on policy decisions affecting online academic work. We can only speak from the perspective of faculty who look in from the outside, who are affected by policy decisions. And what we see is more inconsistency and politics.

Policy decisions concerning Web-based academic offerings seem to suffer from the same sort of equivocality as decisions about the value of online academic work. The university is not sure how to announce online courses, how to charge for them, and how to pay teachers. Most importantly, is the university's policy on the value of online academic work and, hence, the amount of investment warranted. As a nation, we have made a national commitment to Internet education (Year 2000), yet there is not enough hard evidence that the contribution to education warrants such expense (Phipps & Merisotis, 1999). Therefore, decisions about online instruction are made without hard evidence. Social science and research findings are largely irrelevant to the actual concerns of policy makers; instead, what is relevant, but often overlooked, is how ideological factors mediate the policy-making process (Miller & Fredericks, 2000). There seems to be a bandwagon effect, as if universities are thinking: "We have to do this. We can't be the only college without computers, without online offerings—the students won't come here." But such policy precedes evidence to warrant the large investment of limited resources. So, while colleges increasingly give over larger and larger parts of their budgets to online access, they continue to devalue the academic work itself.

Academic Collaboration

Students have used e-mail largely to encourage one another academically and personally, which means that e-mail can provide an excellent means of collaboration (Bonebrake, 2002; Doherty & Mayer, 2003; Khan, 2000; Kwok-Chi, 2001; Zagorsky, 1997). E-mail also may have positive effects on social behaviors such as collaboration and motivation (Tao, 1995). Communicating and collaborating for scholars starts well before the point where they interact with online journals. It takes place between people, and this happens in a great variety of ways. Some academic colleagues who used to talk to one another at the annual conference and rarely after that, now are in frequent contact over the Internet. They can chat and catch up on family matters and engage in co-authored writing or consult one another and provide support and feedback. People who have never met face-to-face may engage in online collaboration. Comeaux (2002) provided an anthology of writing about academic collaboration via the Internet.

Both online journals and paper journals can facilitate the peer review process by using the power of the Internet to communicate with authors and reviewers. For instance, the journal, Communication Education, has a number of ways to communicate with authors and reviewers built into their site.[13] Prospective authors will find a call for manuscripts, indicating what the journal is looking for, a statement of editorial policy, submission guidelines, and even a listing of the editorial board members. Reviewers will find a form used in reviewing manuscripts. Authors can track the progress in reviews of their manuscripts, as they wait to find out the decision of the peer review process. All this Internet information is a huge improvement on the way in which communication typically occurs through snail-mail-only correspondence.

The Internet can intensify the interplay of students who collaborate with one another as they work through a course, whether it is a course entirely given online or supported by online discussion. For students, discussion is a large part of the collaborative process, so students can work through the availability of online discussion boards and small groups online, which are often built into courseware that tracks student online activity. Thus, student collaboration is facilitated by computer-mediated communication. Our students tell us that they find such courses require much reading and writing, and online no one can "hide" and everyone must participate.

Faculty colleagues may provide support to such discussion boards by entering into the discussion from any distance. Online live panels are doable with a bit of planning and adjusting to time zone differences. For instance, we designed an online panel for a course offered in Maine, with two faculty members live online from Karlsruhe, Germany, one from Missoula, Montana, and one online from Kansas City, Missouri. The students sat in a classroom with a computer and LCD projector displaying the discussion board on a movie screen at the front of the room. The class engaged in discussion with these four faculty members with their professor entering the text. Each panelist had provided a brief biography and a question to think about in advance of the panel session. With the exception of a few hitches, it was a grand experience for all. The possibilities are limited only by our imagination. For the most part, faculty are only beginning to explore this realm of communication on the Internet.

In the past, undergraduate students were consumers of scholarship but not producers, except in those rare instances where an outstanding student wrote a conference paper or published a paper in print. However, with the Internet, undergraduates are able to take part in contributing online scholarship. The web offers tools that are accessible and free, which encourage collaboration (Bell, 2001). Users can designate "friends" who are able to access each other's stored files, a tool for encouraging student communication outside the classroom. "Students can conduct joint research or prepare for presentations by sharing files and creating projects together. Any designated file can be accessed by any student, updated, and put back into storage. Several students can work on the same file, although not simultaneously" (Bell, p. 32). In addition to online storage services, the Internet Bookmark Manager makes it possible to access web sites, using a bookmark list available on any computer you are using. Further web page capture utilities allow the user to store web pages for retrieval at a later time. The Internet Bookmark Manager is another tool that the online scholar may use to build up resources in preparing to construct a document.

But even with simple online discussion boards or e-mail, students are finding that collaboration is a valuable form of communication and scholarship that they are likely to encounter as they use the Internet. Because the world of business and the world of academia are using the Internet for collaboration, students may find they are required to practice this form of communication. Students are not alone. Faculty too are finding that the Internet makes it more likely that they will collaborate with colleagues at a distance, which may improve the quality of scholarship available to everyone.

Reflection

Why are there Internet paradoxes in education? The pioneer status of the online academician may account for the controversy. Pioneers have been recognized for both extending the frontiers of Western culture and "at the same time, leaving less space for the other peoples of the world to stand" (Bohannan & Plog, 1967). Viewing the great beyond, thanks to our pioneers, has also had the reverse effect of allowing us to see our institutions and ourselves more clearly. This experience is not always comfortable or comforting. Changing the boundaries in human interplay is likely to generate controversy.

A second cause of the controversies may be a poorly developed theory of online teaching. And the politics? The politics come from this ignorance. The forces of wanting this or that, the power struggles, follow from the differing conceptions of how online teaching works and what it will do.

A third explanation is simply this: Online teaching and learning is neither good nor bad. It is more akin to speaking of blackboard-teaching, or book-teaching, or small-group-teaching, or lecture-teaching. The Internet is a tool. In the hands of a craftsperson, the tool works very well. In the hands of a novice, it is clunky. Some see the finer, more effective uses, either in reality or in imagination, while others see the ineffective uses. Neither perception tells the whole story, as some research is beginning to support.

Consider, for example, the 1996 National Assessment of Educational Progress study of school computers and student achievement scores in mathematics. That study of fourth and eighth graders from a national database of student scores showed that after factoring out students' socioeconomic status, class size, and teacher qualifications, certain kinds of technology use produced higher achievement scores, but other kinds of technology use were associated with reduced achievement scores. When computers were used for higher-order thinking, students performed better, but when they were used largely for drill and practice, students performed worse (Cox, 1997). The conclusion: It depends on how you use the tool. Cox offered a very different explanation of how we conceive of and try to manage new communication technology. In his view, we are mistaken to think that we can engineer and manipulate human communication systems. Instead, such systems evolve as people interact with the medium:

> The manufacturing age's infatuation with central planning and design is still very much with us today. Engineers, scientists, politicians, managers, and educational theorists still speak as if we can, should and therefore must design unimaginably complicated human systems as if they were steam engines or automobiles, manipulating them externally as if the designer were not an integral component of the human process that produces them. But what actually governs complex systems is rarely the industrial age's notion of design at all. Rather, they evolve, shaped by an interaction in which system and environment minutely adjust to each other as biological organisms evolve within ecologies. (Cox, 1997)

Into this array of ideas about online education and misconceptions about how it works and what goes into it, add people who are struggling over power, trying to find the resources they want. There we have the politics. Is it any wonder that there are contradictions in online academic work?

Educators and administrators have different goals. Often the teacher is concerned with the quality of the course and not the expense involved; the administrator, on the other hand, is often concerned primarily with expenses. Hence, Internet teaching may mean more individualized emphasis to the teacher but more students (tuition dollars) to the administrator. We are still struggling with issues of legitimacy of knowledge in this new form. In a special issue of Telecommunications Policy Online, Anthony Rutkowski introduces the question of the future of the Internet: Next-generation Internet (NGI):

> The Internet itself has become a combination incentive and enabling mechanism for an almost incomprehensible number of autonomous development and entrepreneurial activities that constantly come into existence and have global reach and effect. It has become its own self-sustaining means of re-invention. Ultimately, whatever emerges will be some kind of constantly evolving, grand recursive concatenation of all the myriad next-generation developments within developments. (Rutkowski, 2000)

What does research say about the computer and Internet and education? If you investigated online databases, you could find thousands of sources on this topic. There is no way to tell you everything you might want to know about the topic in this brief book. Our objective is to give you ideas about Internet paradoxes in the educational context (see Table 11.1).

TABLE 11.1 *Internet Paradoxes in Educational Contexts*

More educational freedom.	Less educational freedom.
More work and effort.	Perceived as less work and effort.
Desire to save money, time, and resources.	Spend and waste money, time, and resources.

For students, Internet-mediated communication seems to have many positive effects. McCormick and McCormick (1992) found that e-mail of college students served a primarily social function, with about 25 percent of messages containing intimate content. Few messages conveyed hostility or social inappropriateness. In fact, the Internet has enabled students and teachers to do everything they do in the classroom and more. Or has it? Does the Internet facilitate your learning or do you think you'd be better off without computer-mediated communication? The research consistently shows learning online to be as effective as learning off line. But controversies surround the pros and cons of the online method. You may want to conduct your own research about the effectiveness of learning online by using ERIC or Expanded Academic Index. You will find that certain aspects of learning may be intensified while other aspects seem to be missing.

Perhaps the paradoxes in mindsets over Internet territory derives from our old conceptual maps. The paradoxes we witness may reflect our contradictory beliefs about online academic work. When all is said and done, the core of the educational process does not change simply because the material is online—we still need to read critically and write well. How does the Internet intensify your interplay in the learning process? Does the Internet lead to a better understanding of the content you are trying to learn? Process and content still remain two distinct and significant components in human communication. We cannot mistake one for the other.

Case for Discussion

In your course discussion group, give your analysis and perceptions on the following. Internet classes for senior citizens are a popular form of lifelong learning. The more educated a person, the more likely she or he will be connected to the Internet and the more likely they will be interested in learning online, either informally or formally (Burwell, 2001).

1. Discuss how the Internet can contribute to life-long learning.
2. Why do you think Internet learning is so popular among elders?

Internet Investigation

1. *Interview Older User.* Talk to a friend or family member—who is an Internet user and is at least fifty-five years-old—about how she or he uses the Internet to learn. Explore the web site www.aarp.org/expedition/ and ask your friend or relative about the educational opportunities offered there.
2. *Interview Educator.* Contact a teacher who doesn't like to use the Internet about why he or she avoids using the Internet for educational purposes.
3. *Plagiarism.* Investigate various sites designed to provide plagiarized materials and to prevent plagiarism (e.g., turnitin.com). How can students be confident they are using the Internet and database information appropriately? How can educators be confident they are receiving original work from students?

Concepts for Analysis

We have provided concepts relevant to this section so that you can check your understanding. These concepts are worth dissecting in more detail. As a research option, examine one concept. Use our citations as a starting point for your study. We recommend that you search for refereed articles in an online database. Find at least three solid references you can read. Summarize, scrutinize, and present your analysis to other people in your course.

Questions to stimulate your inquiry: Do you agree with the ideas as we presented them? What other points of view should be considered? What details need to be added to fully understand the concept? What can you contribute through your personal analysis of the concept? What more do scholars need to know about the topic?

Distance education is when students take courses away from the campus through mediated communication, such as videos, television, and the Internet.

ERIC is the Educational Resources Information Center, which is a government repository of information about education and related fields. ERIC is considered by some people to be the best database of articles for the field of communication. See http://ericir.syr.edu/

Expanded Academic Index is an excellent full-text database, which contains more than 500,000 citations to articles, news reports, editorials, biographies, short stories, poetry, and reviews appearing in some 1,500 periodicals. See http://www.galegroup.com/tlist/sb5019.html

Online education refers to course work done on the Internet, using a web site, searching the Internet, using e-mail, a discussion board or a course environment. Online education may be a supplement to a face-to-face course.

Outcomes-based education is a results orientation.

Preservice teacher education is instruction designed to help faculty learn to be effective teachers. This faculty development trend often includes infusing technology.

Student as consumer is the approach where schools view the student as a consumer to be satisfied, so educators give students what they want, regardless of what faculty believe constitutes appropriate learning.

Supportive online education helps conventional instruction through web pages, e-mail, and mailing lists, course environments and other means (Hanson & Jubeck, 1999).

Notes

13. http://www.wsu.edu/~comedj/

Part V

Implications

This final chapter is unique in that we attempt to point you toward the implications of communication on the Internet. We summarize ideas we have discussed throughout the book and ask you to contemplate future directions.

12

Consequences and Conclusions

In 1969, when few commercial communications networks existed, a U.S. Defense Department research agency created an experimental system that would eventually become the Internet. Driven by both research and military considerations, the designers of the Internet created a complex, robust, and flexible system that differed in significant ways from contemporary commercial communications networks. In the 1970s and 1980s, computer manufacturers (mainly based in the United States) and telecommunications carriers (mainly operating outside the United States) vied to offer commercial network products and services, but no single company or technology was able to dominate the market, in part because computer users preferred the type of nonproprietary technical standards used in the Internet. In the 1980s, the National Science Foundation *took over operation of the Internet, and in the 1990s the NSF turned over the network to private-sector operators. While the Internet has rapidly increased in scale under commercial ownership, the technology also continues to reflect the system's research origins . . . .*

The transition from public to private operation of the Internet in 1995 brought changes for both providers and users of the service. Commercialization of the Internet allowed ARPA and NSF to get out of the business of running a national network and focus on their research missions. It gave universities, corporations, and government agencies the option of purchasing a standard and reasonably priced Internet service, rather than having to operate their own networks. And it offered online content and communications to thousands of ordinary people who otherwise would not have had access to networks at all.

—(Abbate, 2001, para. 1, Concl. 1)

Today, millions of people in the United States can communicate virtually. We have an accessibility that is not present in all the Americas, and for that matter globally. But the combination of business and nonprofit approaches in the United States has created a tension. In its early days, the Internet represented access to information and discussion on a grand scale, and the opportunity existed to voice one's ideas and to freely seek information. With the commercialization of the Internet, however, we see a striking move in the direction of controlling the flow of information and discussion. That tension has prompted new innovations and a varied approach, which could yield an Internet, which is flexible, adaptable, fast, open, changeable, and a universally usable system. But, depending on how it develops, it could also leave people out. That Internet system will continue to be integrated into other electronic means of communication designed to enhance the services and capabilities that are already present.

The ideology displayed in the early days of the Internet represents a longing in America for certain values, which Fernback (1997) describes as "virtual ideology." She writes:

> There is a "virtual ideology" in cyberspace which is collectivist in orientation. There is a strong sense among users that, despite the tolerance needed for the space to be open-minded and despite the potential for oppressiveness, virtual interaction gives users back some of their humanity—a humanity which is authentically expressed among its constituents via a mass medium whose content is not wholly determined by corporate executives. It is an ideology that characterizes collectivist rhetoric as something positive, not something anti-American or anti-democratic. (p. 46)

Just how this tension is played out is crucial for Americans and all people. It sends a powerful statement about what takes priority at the highest levels of human organization. It is what we call a metamessage, a message at the level of the attitude of the sender toward the receiver (and/or toward what is the content of the message—the message). Such metamessages are just as powerful if not more powerful than the content of the messages uttered.

Throughout this book we have pointed to numerous paradoxical tensions that grow out of communication on the Internet. The tension between an open and a controlled Internet is just one. Each tension, each contradiction, each paradox, each opposition supplies grist for the implication mill: Which way will the tension be resolved? Or how will the tension be resolved, if at all?

Focus

```
TO: Students
FROM: paradoxdoc@hotmail.com
RE: Overview: Implications
```

Lenny and Joan review paradoxes about communication on the Internet and suggest implications about ideas presented in their book. They begin by reviewing the concept of convergence. They ask you to investigate technological innovations while contemplating communication perspectives of the Internet.

Think about the communication process as you revisit a model of communication. Although speculation about the future may seem far-fetched, Lenny and Joan suggest two expectations: (a) computers and the Internet will become more capable, and (b) computers and the Internet will become more pervasive. Technological advances should include increased capacity, more convergence, smaller size, greater mobility, increased flexibility, and lower cost. Those capabilities probably will result in more widespread use and therefore, a more pervasive impact on human communication. In a real sense, the Internet will develop into whatever people want. The capabilities and human desires will determine Internet advances.

As you reflect on your Internet investigation, Lenny and Joan ask you to consider the wide variety of outcomes with computer and Internet-mediated communication. What do you think of the research findings discussed? How does communication on the Internet intensify intrapersonal collaboration? How does communication on the Internet intensify interpersonal relationships? How does communication on the Internet intensify paradoxical outcomes? How does communication on the Internet intensify human misunderstanding? How does communication on the Internet intensify play? Polarization? Fear?

The many possible configurations in technology are comparable to the many possible effects within the individual, which makes generalizations and predictions about communication behavior a difficult task. The technology is evolving so quickly that by the time scholars figure out what is happening socially, the technology will open the way for new communication behaviors. Do you embrace the changes caused by communication on the Internet? Do you accept the paradoxical nature of communication on the Internet?

In this chapter we will explore implications of human communication on the Internet. At the end of this chapter, you should be able to answer the following questions.

1. What is convergence?
2. What innovations are in the works?
3. What is a communication perspective of the Internet?
4. What are some of the implications of human communication on the Internet?
5. What are some of the paradoxes of human communication on the Internet?

Convergence

Juan is an avid computer user. He likes to use the computer to conduct all his college research online. Juan's university library provides access to an array of databases so that he can read books, journals, magazines, and newspapers online. In addition, he knows how to find answers to his questions via web sites. Yesterday, he saw a news report about a telephone device he wanted to buy, so he found the device online, bought the device with his Visa, and expects it will arrive at his home shortly.

Chris is an elderly person who uses the computer to connect to others. Although Chris is unable to get out much, particularly during bad winter weather, Chris "talks" daily to family and friends online. In addition, Chris belongs to several e-discussion groups for entertainment, connection, and fun.

Dave and Janet met online and married last Fourth of July.

Jessica worked for a financial institution in New York City when she married and relocated to North Carolina because of her new husband. Her employer installed the necessary computerized communication equipment so that she could continue working from her new location. She telecommutes to New York Monday through Friday.

Two cans on a string, Ham transceiver, telephone, birthday card through the mail, household intercom, walkie-talkie, home robotics, augmentative communication systems that provide synthesized speech for nonspeaking people, a red London phone booth with computer terminal, wearable computers, interpreters and printed phone services for the deaf . . . The list goes on as we contemplate how humans like to create and use devices that enhance their ability to communicate with each other. The original purpose of the Internet was to enable government communication in case of emergency. Who could have imagined how people would come to communicate via the Internet? Can you imagine how human creativity will translate this medium of communication in the future?

As we have discussed elsewhere in this book, *convergence* is the integration of various media. Internet users versus nonusers is an over-simplistic approach to analyzing Internet effects. Convergence enables new configurations of the Internet. Computer-mediated communication today requires the melding of many electronic devices. So to understand the communication process, one needs to view integrated media in a holistic way, while analyzing the uniqueness of various components.

> We live at the beginning of the age of convergence. Widely predicted over a decade ago, we are now experiencing the bizarre phenomenon of the marriage, intermarriage and multiple integration of entire flocks of electronic devices.

Convergence is progressing so quickly and strongly, that many analysts see the imminent melding of 10 or 20 electronic devices into one.

It is a frenetic movement, causing some of today's pundits to see one box at home and one pocket device on the road, replacing all the functionality of the cell-phone, telephone, radio, television, computer, VHF radio, short-wave radio, VCR, CD-music player, DVD movie player, CD and DVD ROM, fax and personal digital assistant (PDA). (Geller, 1999, paragraph 1–3)

Will your home computer soon convey visual and sound e-mails from your friends? Maybe. Computers can now recognize and use speech. You now can use a computerized cash register to have pricing information, use an e-ticket at the airport, and bank by Internet. The technology is there to enable people to communicate through speech and video and to use electronic devices for transportation. You've probably seen computer wheelchairs that help an individual get around and communicate. Computers and robotic devices are commonplace in business. Thus, it may not be long before your home computer will understand and respond through speech and video.

At the same time as we suggest possible future scenarios, however, we recognize the great danger in doing just that. If anything, media history has left a long and quite evident trail of mistaken prophecies. Baron (2000) reminds us of the many inaccurate predictions of how new communication technology would replace older ones, as well as mistaken ideas about how people would use the new technology. You may be surprised to learn that such erroneous predictions—that a newer communication technology will replace another, undermine some aspect of f2f civility, or predictions about how people would use the new communication technology—date back at least to the advent of writing in the Western world and continue through the middle ages, with the printed book, and right on through to the seventeenth-century introduction of the newspaper, the mid nineteenth-century introduction of the telegraph, followed by the telephone, radio and television, and technologies of our own day, including the Internet.

Change in communication technology occurs so quickly and with unexpected outcomes, that one would be foolish to predict changes in communication behavior with any confidence (Baron, 2000; Fidler, 1997). Paradoxically, we could also say that change in communication technology goes slowly. Saffo ("Paul Saffo and the 30-Year Rule") has offered the 30-year rule, which states that it takes roughly three decades for change to go from the laboratory to the public's adoption of communication technology. So, change can appear to be fast or slow, if Saffo is right, depending on where we are in the 30-year period we are sampling. And what makes all this even more difficult to figure out is that multiple technologies and other changes in society may converge to influence the rate of change. No doubt we are seeing quite a bit of convergence these days. Having said all that, here are our thoughts about the future.

Communication Perspective

Perhaps as you read this book, you noticed that we devoted chapters to intrapersonal, interpersonal, and group communication, but stopped short at public speaking and mass communication. Computer-mediated public communication includes teleconferencing, online meetings, public archives of e-groups, and web sites. So, with the concept of convergence, the important point is to analyze the Internet through the general perspective of communication.

- "Internet traffic has quadrupled in the past 12 months. Quadrupled." (Golson, 2001, p. 14)
- People who use the Internet probably communicate more with other people because of the Internet (Riphagen & Kanfer, 1997).
- "As users adopt high-speed connections—and there are now more than 12 million in the U.S. with high-speed connection services to their homes—time spent online doubles." (Golson, 2001, p. 14)
- Soon 40 percent of new personal digital assistants and 70 percent of new cell phones will be capable of using wireless technology to access the Internet (Chu, 2001).
- The size of the Internet is expanding at the rate of more than 2 million pages a day (DiMaggio, 2001).
- People who use the Internet feel more connected socially than do nonusers (Hamman, 1999; Katz & Aspden, 1997; Parks & Floyd, 1996; Wellman & Gulia, 1999).
- "Personal use of the Internet continues to climb. In the U.S., people spent 20 percent more time online this year than they did last year. Abroad, the growth is exponential—reaching a milestone this year of nearly half a billion souls online." (Golson, 2001, p. 14)
- The majority of U.S. citizens consider the Internet more essential to them personally than is the television (Golson, 2001, p. 14). In a span of ten years, the Internet has gone from little use in its beginning in the United States during the mid-1990s to an expected full access in the United States by 2005 (DiMaggio, 2001).

These facts and predictions underscore the convergence of computer-mediated communication in the United States today. When society prepares for the future, it has a better opportunity to develop those skills necessary to cope with change. The Internet has brought a variety of changes to which people of the United States have adapted, but we will need to continue to adapt in the future.

Most people in the United States expect computer-mediated communication and the Internet to be an integral part of their work and home life. Just because we expect an Internet convergence, however, does not mean the influence on the personal and cultural aspects of our lives will be wholly positive. The Internet can touch interpersonal communication patterns, particularly with co-workers, family, and friends. Positive results with human communication on the Internet seem most likely to occur in a society that examines and directs Internet associations and effects. And, of course, as we have suggested, the effects of the Internet on countries less powerful than the United States must be monitored and taken into consideration. Unlike the television researchers who waited to see what effect the medium had on U.S. society, Internet researchers can employ their knowledge to shape the future. Scholars expect a highly individualized nature of Internet effects and their knowledge can guide you toward positive ends.

People in the United States have long had a love affair with the computer and the Internet, despite many broken hearts along the way. By studying what is happening during human communication on the Internet, you could furnish information you need. The rapid expansion in Internet use, the interpersonal and societal effects, the money invested, the time used, and the Internet revolution justifies a deep investigation, understanding, and prediction. What have you explored? Understood? Predicted?

Communication is a continuous, interactive, and ongoing process. Any technology that becomes a part of the communication process will affect the nature of the interaction. Logic and reasoning, human interplay, the communication associations with the Internet, problem solving processes, manipulation of others through communication, and improving relationships are some of the issues relevant to the study of human communication on the Internet. High technology has compressed the normal communication process by bringing people closer together.

As we discussed earlier in the book, one of the most important areas of Internet use relates to interpersonal communication. Among the pros and cons of the Internet is the way it changes human interaction. From one perspective, the Internet intensifies the home as the center of learning and interaction, but from another perspective, the Internet can intensify antisocial behavior. While some scholars suggest that the Internet encourages people to interact in something beyond the real world, other scholars believe the Internet may actually increase the interaction between people, and result in more positive communication patterns within a household or business. The Internet may give people personal ways of dealing with an impersonal world. High tech can lead to high touch. Perhaps the Internet will allow users to be liberated in ways that increase their need to embrace other senses and seek human contact.

An example of using the Internet to improve communication is its use in e-mails to friends. By adapting a composite letter to each individual, for example, the sender can use word processing to communicate more easily and effectively with loved ones. The word processor can help improve the writing style, correct spelling, and decrease mistakes for the sender. Although, we hasten to add that e-mail users have not flocked to using these computerized helpers, and some, in fact, predict that e-mail may have the opposite effect on writing, making it more like casual e-mails (Baron, 2000). As for a composite e-mail letter, perhaps it enhances the communication, or perhaps the receivers feel slighted by receiving a form letter via e-mail. Have you received holiday greetings from old friends that contained an e-mail address? Has the e-mail contact revived the sense of immediacy in your relationship? Recognizing that any predictions are likely to be wrong, it appears to us that the Internet will encourage a continued move in the direction of a casual or informal style of writing, that this will carry over into non-Internet writing, and that writing will increasingly mimic the spoken, conversational form. Is this a good or a bad thing? Will it matter? That is hard to say.

A classic study of the diffusion of home computers found that interpersonal communication was extremely important in the decision making phase of buying a computer (Rogers, Daley, & Wu, 1983). Potential buyers talked to a variety of people, and were most influenced by their friends in the purchase decision. It appeared that the process of buying a computer can prompt people to interact with many others while seeking information. We must recognize that this study predates the World Wide Web and its use by large numbers of people to gain information from one another, including information about computers. But some things may not have changed. People still seek out the opinions of others, just that since the WWW, more of us do that online. Moreover, the computer changed communication patterns after the purchase because people need computer advice. People like to show off their new computer equipment. Today, more of us do that online than in the days prior to the WWW. In other words, interpersonal communication is still an important part of the picture. We know a computer expert who is able to maintain a wide circle of friends because everyone wants to be his friend—for help, advice, and information. The computer expert is well liked and frequently invited to dinner by his associates. Thus, the sheer process of owning a computer can increase interaction with others. And the acquisition of Internet expertise and computer equipment can increase interaction with others. What did you receive as a gift or buy yourself to celebrate your birthday or a recent holiday? A CD burner? A smart phone? A broadband connection? A DVD player?

Related to seeking help from others online and writing, is the process of collaborating. Collaboration over the Internet is likely to be one form of communication that ought to grow. There are a variety of reasons for this prediction. First, we might simply be biased by our own experience in collaborating on writing this book and other papers collaboratively via the Internet. Second, we see that many others report enthusiastically about their experience of collaborating on academic work (Comeaux, 2002). Third, an Internet search for "collaboration" takes you to numerous education uses of collaboration, including K–12, as well as non-profit organizations of many kinds, and business-related discussions of collaboration. Team work via the Internet is widespread at this writing. We think that is likely to grow. Fourth, scholars of organizational communication have made it abundantly clear that the modern day organization is not only making extensive use of teams and virtual teams, but the new ways of organizing and communicating over time and space increasingly influence the very nature of the organization itself, its style of management, structure, work, and communication processes (Shockley-Zalabak, 2002).

Another prediction we will venture, and this one a bit more abstract than above, is the idea that the Internet is transformative in our thinking. It engages our imagination with the idea of a 'space' for multiple purposes, one that is accessed from multiple locations, in more and more ways, and in multiple times, one that continues to expand and to integrate far flung functions, one that blurs boundaries and creates new categories. It would seem a small step to go from acquiring new concepts—new ways of looking at things—to applying these concepts in new parts of our lives. We have seen hints of this in interpersonal relationships changed in the offline world by ideas acquired online; we have seen hints of this in the intrapersonal world, where the construction of self is influenced by our online experiences; we have seen hints of this in the educational domain, where the idea of what a teacher is and does is influenced by Internet processes of communicating; we have seen hints of this is the intercultural world and the cyber public sphere, where political, economic, cultural and social relationships are reconfigured. Recall that at the outset of this book we threw out the idea that a generative idea oftentimes underlies a new technology, a mythic consciousness or cultural energy. Perhaps it is this new space that is at once private and public, intrapersonal and transpersonal that fires our imagination and carries us into the future. Perhaps we are intrigued by a transformative concept that has captured our imagination and motivates us to explore, to see where it takes us. Likely, the intensity of the online experience is tied to the idea of exploring new concepts. If there is a central idea, we are not sure what to call this new concept. Perhaps you could take a try at naming it and discussing your choice on your course discussion board. At one level, some would say that this new idea is the digital language that underlies electronic communication media.

We think it is more than that. The language is necessary but not sufficient to bring about the changes we observe and predict. While we haven't exactly put our finger on it, we think the central or generative idea has something to do with breaking old communication boundaries and creating new ones.

Internet metaphors may serve as a human communication model. There are many direct similarities between components of Internet-mediated communication and face-to-face communication (see Table 12.1). No one seems to question that the phenomenal expansion of the Internet will continue to multiply in future years. Very recently, colleagues tried to argue with us about whether or not we should teach courses via the Internet, which seemed to us to be an irrelevant question. Using the Internet is a given, the question is: How do we use the Internet to communicate effectively? We are on the verge of having Internet access as commonplace in the U.S. home as the television. Children and adults are already using the Internet to understand the world around them. Generations already exist that have grown up knowing computers and the Internet in the home before they developed extensive social relationships outside the family. Most likely, computer and Internet development will be so astounding over the next fifteen years that many of the ideas we propose in this book will seem like common sense or be tossed aside as part of an older generation who never really understood computers.

TABLE 12.1   *Similarities between Internet-Mediated Communication and Face-to-Face Communication*

Internet Communication Process	*Face-to-Face Communication Process*
User	Source or sender
Program or another person	Self or other as receiver
Data of message	Message
Wires (fiber optics) or wireless carrier on the Internet	Channels
Input devices, keyboard, commands	Encoding
Program bugs, interference, slow modem speed	Noise
System location	Communication context or situation
Power surge	Selective retention
Subroutines	Communication patterns
Physical interference	Physiological influences
Computer generations	Human development
Modem hook-up to Internet	Interpersonal communication
Connect	Attention
Crashing	Communication breakdown
Computer language	Human language
Electrical, optical, or waves used in system	Active process
Multiple interactive programs	Multiple receivers

In the Works

Futurists do not make predictions to tell us what will happen tomorrow, but to expand our thinking so that we can prepare to adapt to tomorrow. If we examine briefly the developments that have occurred during the lifetime of today's elderly U.S. citizens, for example, it excites the imagination: The invention of the automobile, television, flight, space travel, satellites, computers, genetic engineering, medical advances, and more. There are people alive today who have had to adapt to many powerful innovative developments.

While we may seem to have done everything, we will continue to advance farther. Many futurists are not too successful in their predictions because they think linearly. In fact, innovation doesn't move in a straight line. Current capabilities indicate massive influence through the Internet in personal life, education, and business. If and when artificial intelligence becomes a reality, the potential influence appears limitless. The whole concept of artificial intelligence raises new questions about who humans will communicate with in the future. Some computers already are capable of learning from experience and improving ability. If computers achieve the ability to think, truly think, then it is conceivable that people could literally communicate with computer systems located in the home or on the Internet. Until now, humans have communicated only to other living forms. Is it possible that the Internet will allow humans to actually communicate with non-lifeforms for the first time in history? Such a development could have a profound impact upon all human communication. The Internet raises questions about the computer's role in the communication process.

"Computers will seem like dinosaurs in the future, when they'll basically disappear. Wireless Web access will trail you like a shadow. No awkward input devices, either: Just speak clearly" ("The Future Me," 2001, para. 1). Consider these ideas about the future of technology:

- The human body creates heat and energy. Could that be harnessed to power on-body technology?
- Video capabilities on your handheld computer?
- Everyone will telecommute?
- While on vacation, point and click at the Statue of Liberty to download information on the old girl to your handheld computer.
- Dick Tracy's wrist phone? It's in the works complete with a video display, but they can't figure out how to power the thing.
- Body pack. Wear a computer pack and plug yourself into a docking station at school, work, or home? IBM is working on that one.
- E-huge-book? This technology is under development with several companies. Although it looks like a book, it contains hundreds of novels, reference books, and textbooks. Pages are thin, flexible displays that are easy on the eyes. You can still cuddle up with your hundreds of books in your hand in front of the fireplace on a cold night.
- Lasik surgery, but you still need glasses? These glasses aren't designed to make you see better, they help you understand better. That is, the computer translates language onto a screen that appears to be eyeglasses. The glasses automatically give language

translations in front of your eye so that you can have eye contact with that person from another land while reading what she or he is saying on the eyeglasses. IBM is working on this idea.
- "ArrayComm, based in San Jose, California, is pushing a high-speed wireless data system called iBurst, which relies on 'smart antennas' to make more efficient use of radio spectrum than 3G networks. Time Domain, based in Huntsville, Alabama, has developed a system called 'ultra-wideband' transmission that uses low-power radio pulses to transmit data at high speeds over short distances. Even now, more than a century after Guglielmo Marconi pioneered the transmission of wireless data, there is clearly vast scope for further innovation." ("A mobile future," 2001, para. 8)
- The ultimate shoe to "Just Do It." You know how the coach is always trying to teach you the right moves? Well, soon your IT shoes could be able to record your anatomical movement information and give that to your coach, physical therapist, or physician for analysis ("The Future Me," 2001).
- Handheld computers with Internet service are a new alternative that offers advantages to users by providing information and e-mail access (Wildstrom, 2001). There is even clothing designed for you to carry all your techno-gadgets in a 15-pocketed fisherman's-style vest with Velcro-secure conduits that connect wires from pocket to pocket (Wildstrom, 2001).
- No business cards? No problem. Just point your wireless Internet at my wireless, and transfer the information you want me to have. Also called smart phones, these handheld communication media provide Internet access, cell phone capability, address book, calendar, and notepad.
- The prototype of an Internet car was built by Bunnyfoot Products ("Bunny in the Headlights," 2001). The car is equipped with a sound system so the driver can listen to the Internet speak instead of trying to read a screen. The connection uses mobile phone technology.
- Among the wireless systems in the works are V Mail, which is a writing pen that also records the human voice and sends voice mail 100 feet. Lightening mail has a small LCD screen and keyboard that sends wireless messages 50 feet or anywhere when connected to a phone jack. Send It is a data system that records and reads messages on plastic cards (Mariott, 2000).

> "Are you wired?"
> "No, I'm e-smart."
> "Are you online?"
> "No, I'm pulsed."
> "Are you plugged in?"
> "No, I'm waved."

The longer-term effect of existing technology is hard enough to imagine as it is, but there are even stranger things in the pipeline. One trend to look out for is ad hoc networking, in which the network architecture is much more fluid than in today's systems. Handsets might, for example, double as portable base-stations, routing data to and from other nearby handsets. So far the best working example of this idea is Cybiko, a Russian-made electronic toy that allows users to exchange messages and play games with other users nearby. Teenagers lap it up. Jens Zander, of the Royal Institute of Technology in Sweden, imagines a mobile network where each user is given a six-pack of small base-stations when he buys a handset, and is asked to sprinkle them around randomly; the network has no centre, and grows virally. This sort of idea is reminiscent of the "peer-to-peer" approach used by Napster, the infamous music-swapping service. It also resonates strongly with those building guerrilla Wi-Fi networks. ("A mobile future," 2001, para. 7)

Speculation about the future may seem far-fetched, but at least two ideas are overwhelmingly likely to occur: (a) computers and the Internet will become more capable, and (b) computers and the Internet will become more pervasive. The capability should include increased capacity, more convergence, smaller size, greater mobility, increased flexibility, and lower cost. That capability should result in more widespread use, and therefore a more pervasive impact on human communication. In a real sense, the Internet will develop into whatever we want. The capabilities and human desires will determine Internet advances.

Reflections

There are a wide variety of outcomes with computer and Internet-mediated communication. This is a good point to list an array of research findings. Among the positives of Internet communication: increased learning, involvement, speed, increased efficiency, individual adaptation, connection to new people and ideas. Because the computer and Internet have an ability to adapt to the individual, the Internet shines in such things as individualized instruction, personal play, helping people, and meeting individual needs. As an instructional tool, the Internet can be useful in drill and practice, tutorial simulations, data management, and individualized course instruction. The Internet enables people to seek new information and ideas and is a tolerant and forgiving teacher. The Internet can serve as a great equalizer. The Internet offers new ways to succeed. The Internet allows children to be as skilled as their parents. The Internet provides an acceptable and productive mode for the social outcast. While the Internet may make people more alike, it may celebrate our differences and it does seem to give us opportunities to be more equal. Computer- and Internet-mediated communication can increase efficiency, productivity, and safety from dangerous operations. The Internet may give us more free time to spend with loved ones or in recreation. People may meet with greater success by using computers because of their increased likelihood of seeing a task through. The interesting and absorbing qualities of computers can encourage those interacting with them to do so for long periods of time. Concentration may be easier. The Internet can enable users to save money through everything from increased productivity to better online purchases that help the family budget. The Internet opens the world to the elderly, the very young, women, minorities, people who have disabilities, and developing cultures. The Internet can allow an independence never before experienced. There is the opportunity for new topics of conversation and new opportunities to meet people. The Internet may entice users—particularly those who know how to do programming—to increase their ability to think logically.

That process, in turn, can increase creative problem solving and may eventually lead to a more intelligent society.

There are negative outcomes with computer- and Internet-mediated communication: the isolation of people, computer domination to the neglect of other priorities, domination by some cultures over others, a growing centralized and commercialized control, and connecting with people who reinforce anti-social ideas and behaviors. The Internet separates the haves and the have-nots. It widens socioeconomic gaps by making access to the most important technology easier for some people and more remote for others. The Internet allows work to invade our private lives and leisure time. The computer raises expectations so that our work must be more perfect. We can become so absorbed that we fail to interact in socially effective ways. The computer may increase intercultural clashes as people of varying communication styles attempt to engage one another online. There are poor programs, incompatibility problems, the time and expense of computer-mediated communication, and the resulting frustration and fatigue. Those who claim women are less inclined to think logically, less mathematically, less electronically skilled, can use the Internet as another excuse to exclude women (or any other group they wish to oppress) from resources and success. The Internet, with its high-speed operation, may add to an environment that values efficiency over other priorities. We have little doubt that the paradoxes of the Internet can be extremely individualized and can blossom into full-fledged effects, one way or the other.

This book examined the nature and influence of Internet-mediated communication. The Internet is a social and psychological activity, not simply a body of data on a technical medium. Do you think the principles are generalizable? Or do they only hold for specific, individual cases?

The Internet enables the self to communicate intrapersonally (Figure 12.1). The Internet can empower the self through access to information, sheer self-expression, and access to new people and ideas. Although intrapersonal communication and the Internet may seem worlds apart, working online intensifies understandings, organization, and structure of one's intrapersonal communication. Following this line of reasoning, we hypothesize that the nature of communication will be increasingly understood in intrapersonal terms, more and more as a cognitively based model. Closely tied to the intrapersonal model of communication is a tendency to explore the subtlety of meaning. As we view the commu-

FIGURE 12.1 *Communication on the Internet Intensifies Intrapersonal Communication*

- Empowers the self through support and structured processing.
- Enables learning and understanding.
- Creates new ways in which knowledge is created, expressed, and distributed.
- Creates new freedoms.
- Creates new ways that we see ourselves.
- Requires us to integrate information at a distance into local ways of thinking.

nication process as, at its center, an intrapersonal process, with a focus on meaning, we will attend more and more to subjective experience and ethnographic methods of research.

The extensive resources and databases on the Internet provide potential for amazing speed, accuracy, and access to information. The Internet can help advance knowledge for the individual and for civilization because the Internet intensifies our research capabilities. As a reflection of our democratic society, the Internet can offer information, freedom of expression, and education to anyone who has access to the Internet. The educational world opens to each individual who has access to the Internet. The individual can solve problems, find information, search out people and resources on the Internet. The speed of careful, scholarly work is staggering compared to what was possible twenty or even just ten years ago. Today, through the Internet, a scholar can accomplish in a matter of hours what used to take days or weeks of effort. Databases permit quality searching to find information, while many databases provide the full text of articles the scholar needs to read.

> The star of the current conflict in Afghanistan is the video satellite phone. This allows live broadcasts to be beamed directly from a war-torn and technologically primitive country. Liberated from landlines and truckloads of satellite dishes—the modern equivalent of those horse-drawn carts—journalists can talk to the camera where they please, without being beholden to the authorities ("Science and technology," 2001).

The world has become a global village through the technology that makes us aware of the people of the entire planet. But as we pointed out, Western culture is dominating this new Internet medium. Although people have many different uses for the Internet, communicating with friends and family is a major use.

People use the Internet to meet people and develop new relationships (see Figure 12.2). They use the Internet to enhance relationships with family and friends they already have by keeping in touch, talking about day-to-day activities. The Internet can be especially helpful in maintaining long-distance relationships when colleagues, family, and friends live far away.

Does the Internet democratize and equalize the U.S. society or cause a digital divide? The Internet can serve as a mediator of human communication and a way to develop community. Many online discussion groups provide members with information, but they also can provide participants with a sense of community and belonging. One advantage of groups is that people who may feel marginalized in our society—e.g., women, minorities,

---

FIGURE 12.2 *Communication on the Internet Intensifies Interpersonal Relationships*

- Connect to family and friends.
- Provide opportunity for storytelling.
- Create new relationships.
- Maintain better relationships.
- Bring people together.
- Contribute to the development of community through e-groups.
- Empower people through support groups.

people with disabilities—may be able to connect with people online in unique ways. For example, according to McKenna and Bargh (1998), one advantage of computer group memberships is that gay men and lesbians who might be marginalized in other contexts can belong to groups that become a part of their identity. The relatively anonymous nature of online groups gives participants a place for interaction that they might not have otherwise.

Some groups have explicit rules of behavior and others have implied expectations. Online groups function with the same kinds of social rules that one encounters in any group. Members who participate successfully gain greater self-acceptance and learn skills that can enhance their other social interactions (McKenna & Bargh, 1998). Other topics of relevance to group communication were high volume e-mail, Internet anonymity, virtual reality, lack of nonverbals, and considerations for moderating online discussion. You also saw how the Internet can provide a means of advocacy, by looking at the example of health care empowerment.

Computer systems, sociotechnical factors, time, hardware, and software affect the way people communicate individually on the Internet. If you communicate with a friend via America Online (AOL) Instant messaging, for example, you can interact differently than on a Blackboard or WebCT message board. There may be differences, for example in how the communication appears visually, time synchronization, interactivity, and privacy. In instant messaging you may not give the same kind of thought and effort you give in preparing your message on Blackboard or WebCT. The IM will be brief, may have no punctuation, use nonstandard capitalization, and show greater informality. Your IM may take on a rhythm of banter or interplay back and forth. The message posting in Blackboard or WebCT may take a more one-way, longer, and more formal style. This messageboard style is probably still shorter and more casual than if you wrote a snail mail (U.S. Postal Service) letter. Thus, not only does your individual perception (intrapersonal communication) affect your communication, but the technology that you use can affect the way you communicate (and perceive) on the Internet.

Whether you are an avid Internet user or a novice, your individuality affects how you use the Internet. And your particular Internet use may affect other elements of your social interaction. Sometimes heavy users are made to feel defensive, as if they are being antisocial. But Internet users are probably at least as social as anyone else (Robinson, Kestnbaum, Neustadtl, Alvarez, 2000). In fact, Internet users "read more literature, attended more arts events, went to more movies, and watched and played more sports than comparable nonusers" (Robinson & Kestnbaum, 1999). Paradoxically, while a problem for some people, for the average person, the Internet enhances relationships and cognitive and social interplay. We cannot say whether these characteristics are part of the makeup of extensive Internet users or if the Internet creates this effect, but several studies suggest that Internet users have more trust in general and bigger social networks than nonusers of the Internet (Cole, 2000; DiMaggio, Hargittai, Neuman, & Robinson, 2001; Uslaner, 2001; Wright & Bell, 2003).

One of the paradoxes of the Internet is the discussion about the potential for change (Figure 12.3). Some analysts believe that the Internet has created profound changes: "The new economy is about communication, deep and wide. All the transformations . . . stem from the fundamental way we are revolutionizing communications" (Kelly, 1998, p. 4). Other people think that the fundamentals remain the same.

FIGURE 12.3 *Communication on the Internet Intensifies Paradoxes*

- Intensifies fears while giving an outlet for facing fear.
- Brings people together while polarizing others.
- Solves educational problems while creating educational controversies.

FIGURE 12.4 *Communication on the Internet Intensifies Emotions*

- Intensifies concentration.
- Intensifies expression of feelings.
- Intensifies hostility and hate.
- Intensifies imagination.
- Intensifies sense of connection.
- Intensifies learning.

While some people make a fortune, others go bankrupt. After the initial euphoria about Internet companies, the NASDAQ fell. After the NASDAQ fell, online business continued to increase. While the Internet can connect workers all over the country who work out of their home, the workers may feel isolated from each other (Figure 12.4).

Because the Internet is a medium of human communication, it shapes that communication. One way of shaping is by strengthening and intensifying certain elements of the interplay between people.

If you are an avid Internet user, you probably have felt surprised when you suddenly realized how much time had passed after you had intently focused on the Internet for awhile. Perhaps you felt a sudden rush of physical awareness after intense Internet use. We think that is because the Internet can intensify your ability to concentrate on a task so that you have strong concentration. The involving elements of the Internet keep your attention focused.

Skill Section: Internet Behavior

The way we attach meaning to symbols is crucial to communication. The Internet confuses meanings because it blurs, creates a gray area, makes participants less clear of meaning (Figure 12.5). Meaning is blurred because boundaries are blurred. Communication on the Internet can lack precision, clarity, and understanding of text because it lacks nonverbals, blurs boundaries between work and the personal, blurs boundaries between fantasy and reality. What are some general behavioral implications for clarifying communication on the Internet?

FIGURE 12.5 *Communication on the Internet Intensifies Misunderstanding*

- Lacks clarity and understanding of text because it lacks nonverbals.
- Blurs boundaries between work and the personal.
- Blurs boundaries between fantasy and reality.

Confirm Important Communications

Despite the advantages of Internet communication, there are limitations. When the communication is important, double-check your perceptions and the way you attach meaning by following up with more communication or a different kind of communication via another medium.

Engage Your Sense of Humor

Take communications with a grain of salt. Know that people may be playing, using role-playing, playing with their identity, or entertaining themselves. The results of an experiment using humor in online interactions suggested that humor may "enhance likability" of the other person. The humor led participants to increase their sociability, cooperate more, and perceive that they were more similar to their work partner. Given that business people tend to reduce social talk with their increase of e-mails, appropriate, context-based humor online may enhance their work communication. We are not suggesting forwarding the endless e-mail jokes, but you may want to find a way to convey your sense of light-heartedness or humor while online in ways that are similar to the ways you would do so in face-to-face interactions (Morkes, Kernal, & Nass, 1999).

Of course, one of the tricks to effective Internet communication is to make sure humor and play are not misinterpreted by the receiver, particularly when there is an expectation of behaving with *professionalism* or all-business in professional contexts. With the importance of a positive climate and good interpersonal relationships in business, employees will want to make sure they continue to find ways to communicate effectively in computer and Internet-mediated communication so that they enhance their work communication.

Ask Questions

Because the lack of nonverbals and the brevity of text delete important information about the meaning of messages, be careful. On important messages, you'll want to make sure you attach meaning appropriately.

Use Netiquette

There are certain rules of behavior expected online. In discussion groups there may be formal and informal rules of behavior. In business contexts, you'll want to use the same professional behavior you'd use in sending a hardcopy letter. E-mail etiquette suggests that the subject line should clearly state the contents. In addition, you should pay attention to grammar, punctuation, and spelling. Avoid attachments, particularly large ones. And avoid forwarding jokes. Use plain text and definitely avoid html format because it can contain a type of virus. Be aware that you may benefit from waiting a day before you send an angry or emotional letter. Some organizations prohibit chain letters and those kind of e-mails because they tie up transmission and server space (Colombo, 2000).

When people interact cognitively or socially, they interplay (Figure 12.6) with the self or other in these ways:

FIGURE 12.6 *Communication on the Internet Intensifies Play*

- Imagination encouraged.
- Role-playing.
- Interplay with the self or other.
- Play that actors act in.
- Play games for destruction.

- *Imaginative play like a child.* Circulating humor, jokes, having fun, and playing Internet games are some of the kinds of Internet play available to users. Some people respond positively to creating virtual experiences online, perhaps because they already use and enjoy the medium for computer games, MOOs, and MUDs. E-discussion has been called a *cyberspace playground* because the interaction is like child's play (Scodari, 1998). Some Internet users have playful entertainment as a goal in the Internet communication.
- *Identity role-playing.* The Internet allows users to experiment with their identity. If you talk online anonymously, if you pretend to have a different job than you really do, you are role-playing through the Internet. Consider the combination of identity role-playing and child-like play suggested by role-playing sites (e.g., www .roleplayinggames.net,http://dir.yahoo.com/Recreation/Games/Role_Playing_Games/).
- *Interplay with the self or other.* Scholars often refer to human communication as human interaction, intercourse, and interplay (e.g., Adler, Rosenfeld, & Proctor, 2001; McIntosh, 1995). In this book, interplay means communication, but we intend to arouse certain connotations with our use of the word interplay. The words interaction, intercourse, and interplay bring slightly different ideas to mind. Precision in language can help clarify meanings, so we intentionally use the word interplay because of the root play. We believe that Internet users approach their communication activity with a sense of playfulness, and users seek ways to use the Internet that is pleasurable.
- *Actors in a play.* Virtual reality is the creation of an imaginary space in which people can pretend. Of course we can question to what extent the play and actors are real or imaginary.
- *Games for destruction.* In the "home" videotape of Bin Laden released by the U.S. government (December 13, 2001), Bin Laden described his feelings of pleasure watching the destruction of the World Trade Center. Bin Laden compared his feelings of elation at killing the people in the World Trade Center with the experience of watching a game of soccer in which his team won. Bin Laden's metaphor suggests that he perceives terrorism as a fun game he is playing. Although this vile extreme is unusual, there are people who play games on the Internet that are designed to destroy the lives or resources of others. The spectrum ranges from gossip to terrorism. Although the destructive effects vary, gossip, virus transmissions, S&M sex sites, and violent videogames often take on a sense of gaming for the Internet user, so that they receive the rush of winning a game that takes something from other people (time, status, money, feelings of safety, power, innocence, control).

FIGURE 12.7 *Communication on the Internet Intensifies Collaboration*

- Connect human talents.
- Resolve the Inner/Outer Dichotomy.
- Provide interactivity.
- Collaboration across age, gender, ethnicity.

Intrapersonal communication also can take place in a more collaborative way. A key use of the Internet is to make work more efficient, of higher quality, and more effective. Collaboration (Figure 12.7) is an essential element of the value of Internet communication in the business context. Internet technology has increased business' ability to have more people involved in decision making and team cooperation that spans the miles.

Computer use affects all people within our society because we have easy access to information and people around the globe. We can use computers to communicate to people who think like us and ones who hate us although they've never met us. Communicating through the computer can help us make or meet our enemies. In other words, the Internet can intensify polarization of people (Figure 12.8). Through the Internet, people who are similar cluster together, reinforcing each other's similarities while standing against people who are different. In some cases, this polarization may be benign, in other cases, the polarization may be part of insidious hate and violence.

When many people replace their local newspaper with information portals on the Internet, they select filters that adapt to their interests and biases. That process enables their access to the world's news to be seen through filters that can intensify particular attributes.

Are we at greater risk because of the Internet? Should we feel fear because of the Internet? Certainly some individuals are more vulnerable because of the Internet. We had a student, for example, who did not want her name or photo put on the web pages for a course. She had worked hard to escape a violent ex-husband. Her ex-husband had been imprisoned for attacking her, and she had subsequently relocated for her safety. The student was afraid her Internet course might enable this violent man to track her down and kill her. Her fears are real.

Many people believe that they are at greater risk since the September 11th attack on the United States. Are we at a greater risk or do we better understand the risks we face in this age of human communication on the Internet?

The Internet can increase and decrease fear (Figure 12.9). We discussed several concepts related to this paradox. By using the Internet, we gain freedom to access information

FIGURE 12.8 *Communication on the Internet Intensifies Polarization*

- Intensify polarization.
- Create cognitive dissonance.
- Vent through metaphors.
- Speak hate.
- Play violent games.

FIGURE 12.9 *Communication on the Internet Intensifies Fear*

- Personal fear and embarrassment over hoaxes, rumors, and myths.
- Fear of becoming addicted to the Internet.
- Fear of others because of hatemongers, hate sites, stalking.
- Cyberterrorism.
- Fear of loss of privacy.

but we give up our individual privacy because of computer tracking information. Internet hoaxes can create fear and system problems. By getting people to pass along the hoax information, the message acts like a virus by tying up transmission and server space. Rumors and myths seem particularly credible on the Internet because of their written form and repetition. There is evidence that a normal individual can become addicted to Internet use to the point of neglecting personal and work responsibilities, and becoming socially isolated (Belsare, Gaffney, & Black, 1997). Although some people may use the Internet in unhealthy ways, the average user is in no danger of becoming addicted to the computer. The Internet fuels fears and the sense of vulnerability because indeed we are vulnerable. The Internet is used to communicate terrorism, enables stalking, creates opportunity for stealing identities, and facilitates hatemongers. In contrast, the speed and access to information and community and support available online can enable people to overcome fears. For some people the Internet intensifies risk and fear, while for other people, the Internet is a way of coping with threats and reducing fear.

Reflection

Have you hugged your computer today? Or did you feel more like throwing it out a window? Human communication on the Internet is awash with paradoxes. Before writing this book, we first began our investigation while looking at how students, faculty, and administrators communicated on the Internet. We observed many contradictory responses and contradictory information as we saw how various people responded in various ways to communication on the Internet. We investigated personal and business contexts, where people continued to contradict each other in their responses to CMC. We realized that the paradoxes of communication on the Internet appear profoundly influential regarding the subjective or individualistic way people communicate via the Internet. We realized that for the moment, there are more questions than generalizable answers.

On any given day, about half of the people of the United States who have Internet access go ahead and connect online. Although not everyone has Internet access, that access is expanding. Schools, businesses, and homes are increasing their Internet connections. Currently, virtually everyone in this country has access to a television and radio. A similar level of Internet accessibility is expected by the year 2005. Although not everyone will have access at home, they will be able to use the Internet through schools and businesses. And anyone who can go to a public library probably can access the Internet (Harrison, 2001).

As you critique the ideas here, consider the concept of convergence and new technological innovations. Computer-mediated communication is by no means a universal medium with one type of computer and connection. The technology hardware and software is extremely complex with many kinds of differences. The many possible configurations in technology are comparable to the many possible differences within the individual, which makes generalizations and predictions about communication behavior a difficult task. Types of Internet communication include voice mail as well as text e-mail, interactive games, databases, information portals, online support groups, discussion forums, instant messaging, wireless and wired connections, desktop video-conferencing, and more. The technology is evolving so quickly that by the time scholars figure out what is happening socially, the technology is opening the way for new communication behaviors.

Do you embrace the changes caused by communication on the Internet? Do you fear the changes caused by communication on the Internet? Probably you can answer both questions with "Yes." Many people can. That is a significant paradox. We cannot predict the future, but we have some questions for you to consider as you prepare.

> How will the Internet intensify, diminish, and shape human communication in the future?
>
> How will the Internet be used in business and political warfare?
>
> How will we resolve concerns about intellectual property on the Internet?
>
> Will we ignore privacy and security concerns because the advantages of Internet communication outweigh what we give up in our privacy?
>
> What new software and hardware developments will affect human communication on the Internet?
>
> How might your communication expectations change because of the Internet?
>
> How does the Internet intensify your interplay with yourself and others?

Case for Discussion

In your course discussion group, give your analysis and perceptions of the following.
One study of college students found that 73 percent of students accessed the Internet at least once a week, and 13 percent of students indicated that their computer use interfered with personal functioning (Sherer, 1997).
What do you think will be the future of Internet use, particularly for students?

Internet Investigation

1. *Product Development.* Conduct an Internet search about new Internet product development. You will want to consider computer innovations by businesses and major research institutions. Post your findings to your course discussion group.
2. *Future Effects.* Speculate on whether the Internet will affect our lives in coming years due to its increasing commercialization, Western bias, and emphasis on speed,

reach, interactivity, anonymity, and regard. Post your thinking to your course discussion group.
3. *Bionic Vision for the Blind.* Investigate some of the types of product development underway right now that would be able to offer life changes for people with disabilities. You might investigate vision, hearing, or mobility technology. What are the implications for convergence of these ideas?
4. *The Law.* Investigate the implications of law on the Internet (Barlow, 1995; Cavazos & Morin, 1994; Rose, 1995; Samoriski, 2002; Spinello, 2000). Because of the breadth of this topic, you'll want to collaborate with other class members or focus on a particular area of Internet law. What are the implications for you as a user? As a communicator?

Concepts for Analysis

We have provided concepts relevant to this chapter so that you can check your understanding. These concepts are worth dissecting in more detail. As a research option, examine one concept. Use our citations as a starting point for your study. We recommend that you search for refereed articles in an online database. Find at least three solid references you can read. Summarize, scrutinize, and present your analysis to other people in your course.

Questions to stimulate your inquiry: Do you agree with the ideas as we presented them? What other points of view should be considered? What details need to be added to fully understand the concept? What can you contribute through your personal analysis of the concept? What more do scholars need to know about the topic?

Advanced Research Projects Agency (ARPA) is a government initiative that worked in the original development of the Internet. See http://www.darpa.mil/

A *CD or DVD burner* is an electronic device that enables the computer user to create their own compact discs or digital video discs.

Convergence is the integration of various computer media.

Cyberspace playground is a playful approach to the Internet and e-discussion.

DSL or cable are high-speech Internet connections.

National Science Foundation (NSF) is a government-funded organization designed to advance science; it was fundamental in the development of the Internet.

A *personal digital assistant* (PDA) is a Palm Pilot or similar device that can contain a calendar, telephone, email, and other communication aids.

A *smart phone* is a combination telephone and email communication device.

References

Abbate, J. (2001, Spring). Government, business, and the making of the Internet. *Business History Review, 75*(1), 147–176.

Abdal-Haqq, I. (1995). Infusing technology into preservice teacher education. *ERIC Digest.* [On-Line]. Available: http://www.ed.gov/databases/ERIC_Digests/ed389699.html.

Ackermann, E., & Hartman, K. (1999). *The information specialist's guide to searching and researching on the Internet and the World Wide Web.* Wilsonville, OR: ABF Content.

Adkins, M. E. (1994). *Computer-mediated communication and interpersonal perceptions.* (ERIC Document Reproduction Service No. ED 332 251)

Adler, R. B., Rosenfeld, L. B., & Proctor II, R. F. (2001). *Interplay: The process of interpersonal communication* (8th ed.). Fort Worth, TX: Harcourt.

Administration on Aging. (1998). *Your source for information on aging.* Washington, DC: U.S. Department of Health and Human Services Available at: http://www.aoa.dhhs.gov/default.htm Accessed May 6, 2003.

Adrianson, L., & Hjelmquist, E. (1999, May). Group processes in solving two problems: Face-to-face and computer-mediated communication. *Behaviour & Information Technology, 18*(3). Accessed via Academic Search Elite.

Adrianson, L. (2001, January). Gender and computer-mediated communication: Group processes in problem solving. *Computers in Human Behavior, 17*(1).

Agre, P. (1994). Net presence. *Computer-Mediated Communication Magazine, 1*(4).

Aitken, J. E. (1985). Microcomputers in the communication process: The literature, a survey and model. *Dissertation Abstracts International, 46*–10, 3004.

Aitken, J. E. (1987). Home computers as tools to strengthen the family. In J. King & P. Freudiger (Eds.), *Building family strengths: A foundation for wellness.* State University: Arkansas State University Press.

Aitken, J. E. (1999). *Internet communities through online discussion groups.* Paper presented at the national meeting of the OSCLG Presentation on Internet Community, Wichita, KS.

Aitken, J. E. (2000). *The new language: The virtual you.* Paper presented at the national meeting of the National Communication Association, Atlanta, GA.

Aitken, J. E., & Shedletsky, L. J. (1998, November). *Teaching intrapersonal communication with the World Wide Web.* Paper presented at Association for the Advancement of Computing in Education, World Conference of the World Wide Web, Internet and Intranet, Orlando, FL.

Althaus, S. L., & Tewksbury, D. (2002, January/March). Patterns of Internet and traditional news media use in a networked community. *Political Communication, 17*(1), 21–46.

Andersen, P., & Guerrero, L. K. (Eds.). (1998). *Handbook of communication and emotion: Research, theory, application, and contexts.* Burlington, MA: Academic Press.

Anderson R. H., Bikson T. K., Law S. A., & Mitchell, B. M. (1995). *Universal access to e-mail-feasibility and societal implications.* Santa Monica, CA: Rand.

Angeli, C., Valanides, N., & Bonk, C. J. (2003, January). Communication in a web-based conferencing system: The quality of computer-mediated interactions. *British Journal of Educational Technology, 34*(1), 31–44.

Arakaki Game, J. (Ed.). (1998, April). Communication, culture, and technology: An Internet interview with James W. Carey. *Journal of Communication Inquiry, 22*(2), 117–130.

Argyris, C. (1990). *Integrating the individual and the organization.* New Brunswick, NJ: Transaction.

Armistead, T., Cho, A., Ichniowski, T., Rubin, D. K., Angelo, W. J., Tuchman, J., & Sawyer, T. (2001, October 15). Nation struggles with issue of protecting infrastructure. *ENR, 247*(16), 11.

Arnold, A. M. (1998, February). Rape in cyberspace: Not just a fantasy. *Off Our Backs, 28*(2), 12–13.

Arquette, T. (2002). *Social discourse, scientific method, and the digital divide: Using the information intelligence quotient (IIQ) to generate a multi-layered empirical analysis of digital division.* Doctoral dissertation, Northwestern University. Available online at: http://www.sla.purdue.edu/people/comm/arquette/dissertation.htm

Arrow, H. (1997, January). Stability, bistability, and instability in small group influence patterns. *Journal of Personality & Social Psychology, 72*(1), 75–86.

Artz, N., Munger, J., & Purdy, W. (1999, Fall). Gender issues in advertising language. *Women and Language, 22*(2), 20–26.

Ashling, J. (2003, February). Online information's wrap-up session. *Information Today, 20*(2), 40.

Ashton, G., Barksdale, K., Rutter, M., & Stephens, E. (1995). *Internet activities: Adventures on the superhighway.* Cincinnati, OH: South-western Educational Publishing.

Asman, M. F. (2001, January–March). Electronic monitoring of employees. *Ohio CPA Journal, 60*(1), 25–28.

Bakhtin, M. M. (1981). *The dialogic imagination.* Austin: University of Texas Press.

Barclay, D. & Wakabayashi, R. (2000, Spring). Hate crimes and the business community. *Diversity Factor, 8*(3) 32–37, para 13.

Barlow, J. P. (1995, December). Property and speech: Who owns what you say in cyberspace? *Communications of the ACM, 38*(12), 19–22.

Barnes, S. B. (2002). *Computer-mediated communication: Human-to-human communication across the Internet.* Boston: Allyn and Bacon.

Barnard, C. (1938). *The functions of the executive.* Cambridge, MA: Harvard University.

Barnatt, C. (1999, April). Apple pie thinking for the wired age? *Human Relations, 52*(4), 521–537.

Barnet, R. J., & Cavanagh, J. (1994). *Global dreams: Imperial corporations and the new world order.* New York: Simon & Schuster.

Baron, N. (2000). *Alphabet to email: How written English evolved and where it's heading.* New York: Routledge.

Barrett, E., & Lally, V. (1999, March). Gender differences in an on-line learning environment. *Journal of Computer Assisted Learning, 15*(1), 48–60.

Barron, A., & Ivers, K. (1996). *The Internet and instruction: Activities and ideas.* Englewood, CO: Libraries Unlimited.

Barron, M., & Kimmel, M. (2000, May). Sexual violence in three pornographic media: Toward a sociological explanation. *The Journal of Sex Research, 2,* 161–168.

Basch, R. (1998). *Researching online for dummies.* Foster City, CA: Idg Books Worldwide.

Basso, E. B. (1992). Contextualization in Kalapalo narratives. In A. Duranti & C. Goodwin (Eds.), *Rethinking context: Language as an interactive phenomenon* (pp. 253–270). Cambridge, UK: Cambridge University Press.

Bauman, R. (1975). Verbal art as performance. *American Anthropologist, 77,* 290–311.

Baxter, L., & Montgomery, B. (1996). *Relating: Dialogues and Dialectics.* New York: Guilford.

Baym, N. (2000). *Tune in, log on: Soaps, fandom, and online community.* Thousand Oaks, CA: Sage Publications.

Beirne, M. (2000, October). Online appointment scheduling takes hold. *Review of Ophthalmology, 7*(10), 29.

Bell, S. (2001, July). Web-based utilities for learning and collaboration in the classroom. *Syllabus 14*(12), 32–35.

Belsare, T. J., Gaffney, G. R., & Black, D. W. (1997, February). Compulsive computer use. *American Journal of Psychiatry, 154*(2), 289–290.

Benokraitis, N. V. (Ed.). (1997). *Subtle sexism: Current practice and prospects for change.* Thousand Oaks, CA: Sage Publications.

Berlo, D. K. (1960). *The process of communication.* New York: Holt, Rinehart and Winston.

Berman, D. (2001, September 17). E-business: The Web at its worst: Pranks turn cruel, rage finds outlets. *Wall Street Journal,* B, B.6.

Berman, J., & Bruening, P. (2001, Spring). Is privacy still possible in the twenty-first century? *Social Research, 68*(1), 306–318.

Berry, W. (1993). *Sex, economy, freedom and community.* New York: Pantheon.

Biggs, S. (2000, August). "Charlotte's Web:" How one woman weaves positive relationships on the net. *Cyberpsychology & Behavior, 3*(4), 655–663.

Bing, J. (1992). Penguins can't fly and women don't count: Language and thought. *Women and Language, 15*(2), 11–14.

Bishop, A. (1991). *The National Research and Education Network (Nren): Update 1991.* (Eric Document Reproduction Service No. Ed 340 390).

Blackwell, R. D., & Stephan, K. (2001). *Customers rule!: Why the e-commerce honeymoon is over and where winning businesses go from here.* New York: Crown Business.

Blake, R., & Mouton, J. S. (1964). *The managerial grid: Key orientations for achieving production through people.* Houston: Gulf Publishing.

Bodker, S. (1997). Computers in mediated human activity. *Mind, Culture, & Activity: An International Journal, 4*(3), 149–158.

Boggs, C., & Jansma, L. L. (1999). Subtle sexism: Current practice and prospects for change. *Journal of Language and Social Psychology, 18*(2), 223–231.

Bohannan, P., & Plog, F. (Eds.). (1967). *Beyond the frontier: Social process and cultural change.* Garden City, NY: The Natural History Press.

Bonebrake, K. (2002, December). College students' Internet use, relationship formation, and personality correlates. *CyberPsychology & Behavior, 5*(6), 551–558.

Bordia, P. (1996, May). Studying verbal interaction on the Internet: The case of rumor transmission research. *Behavior Research Methods, Instruments, & Computers, 28*(2), 149–152.

Bormann, E. G. (1980, Reissued 1989). *Communication theory.* Salem, WI: Sheffield.

Bowen, C. (1999, August 7). Pigeonholing hate groups with the Net. *Editor & Publisher, 132*(32), 24.

Bowen, C. (2000, April 17). Un-goofy-ing those urban legends. *Editor & Publisher, 133*(16), p. 124.

Brandon, D. P., & Hollingshead, A. B. (1999). Collaborative learning and computer-supported groups. *Communication Education, 48,* 109–126.

Breen, R., Lindsay, R., Jenkins, A., & Smith, P. (2001, March). The role of information and communication technologies in a university learning environment. *Studies in Higher Education, 26*(1), 95–115.

Brehm, J. W., & Cohen, A. R. (1962). *Explorations in cognitive dissonance.* New York: Wiley.

Brewer, S. (1998, April). Online connections. *HomePC, 5*(4), 117–120.

Brody, J. E. (2000, May 30). Mighty cyberengines spew health myths. *New York Times, F,* F.8.

Bruner, M. S. (1999, May). An analysis of e-mail communication. *International Journal for Academic Development, 4*(1), 59–63.

Brunyand, J. H. (2001). *Encyclopedia of urban legends.* Santa Barbara, CA: ABC-CLIO.

Bulkeley, D. (2000). The Internet as a collaboration tool. *Design News, 55*(7), W10, W12.

Bull, K. S., Winterowd, C. L., & Kimball, S. L. (1999, Winter). Uses of the Internet: Rural special education materials for teachers and parents. *Rural Special Education Quarterly, 18*(1), 12–23.

Bunny in the headlights. (2001, January). *Financial Management.* UK, p. 29.

Burke, K. (1969). *A grammar of motives.* Berkeley, CA: University of California Press.

Burke, K. (1984). *Permanence and change.* Berkeley, CA: University of California Press.

Burke, K., Aytes, K., Chidambaram, L., & Johnson, J. J. (1999, August). A study of partially distributed work groups: The impact of media, location, and time on perceptions and performance. *Small Group Research, 30*(4), 453–490.

Burwell, L. A. (2001, November). Too old to surf? No way! An Internet course for seniors. *American Libraries, 32*(10), 40–43.

By the numbers. (2002, May). *T+D, 56*(5), 26.

California Department of Education. (2002). Assessment and Accountability. [On-line] Available: *http://www.cde.ca.gov/ac/*

California State Senate. (2002, July). *Joint Committee to Develop a Master Plan for Education—Kindergarten through University.* Available Online: http://WWW.SEN.CA.GOV/masterplan/

Callahan, C. (1999). *A journalist's guide to the Internet: The net as a reporting tool.* Boston: Allyn and Bacon.

Camp, L. J. (1996). We are geeks, and we are not guys: The systers mailing list. In L. Cherny & E. R. Weise (Eds.), *Wired_women: Gender and new realities in cyberspace.* Seattle, WA: Seal.

Camp, T. (1997). The incredible shrinking pipeline. *Communications of the ACM, 40,* 103–111.

Campbell, S., & Neer, M. (2001). The relationship of communication apprehension and interaction involvement to perceptions of computer-mediated communication. *Communication Research Reports, 18*(4), 391–398.

Cardenas, K. H. (2001). Technology in higher education: Issues for the new millennium. In S. Losco & B. L. Fife (Eds.), *Higher education in transition: The challenges of the new millennium.* Westport, CT: Greenwood.

Carroll, J. B. (Ed.). (1956). *Benjamin Lee Whorf: Language, thought and reality.* Cambridge, MA: MIT Press.

Cassell, M. M., Jackson, C., & Cheuvront, B. (1998, January–March). Health communication on the Internet: An effective channel for health behavior change? *Journal of Health Communication, 3*(1), 71–79.

Cattagni, A., & Farris, E. (May, 2001). Internet Access in U.S. Public Schools and Classrooms: 1994–2000. *National Center for Education Statistics* [On-Line serial]. Available: *http://nces.ed.gov/pubsearch/pubsinfo.asp?pubid=2001071*

Cavazos, E. A., & Morin, G. (1994). *Cyberspace and the law.* Cambridge, MA: MIT Press.

CBS Evening News. (2001, December 7). *Evening news television broadcast.* CBS Broadcasting Inc.

Cegala, D. J. (1984). Affective and cognitive manifestations of interaction involvement during unstructured and competitive interactions. *Communication Monographs, 51,* 320–335.

Cegala, D. J., Savage, G. T., Brunner, C., & Conrad, A. B. (1982). An elaboration of the meaning of interaction involvement: Towards the development of a theoretical concept. *Communication Monographs, 49,* 229–245.

Chadwick, S. A. (1999). Teaching virtually via the web: Comparing student performance and attitudes about communication in lecture, virtual web-based, and web-supplemented courses. *The Electronic Journal of Communication,* 9 (1) [Online]. Available: *http://www.cios.org/getfile/Chadwick_v9n199*

Chadwick, S. A., & Callaway Russo, T. (2002). Virtual visiting professors: Communicative, pedagogical, and technological collaboration. In P. Comeaux (Ed.). *Communication and collaboration in the online classroom: Examples and applications.* Bolton, MA: Anker Publishing.

Cherny, L. M. (1996). The MUD register: Conversational modes of action in a text-based virtual reality (Internet, Linguistic Register). Doctoral thesis, Stanford University. *Dissertation Abstracts International, 56-1,* Section: A, p. 4746.

Chernyshenko, O. S., Miner, A. G., Baumann, M. R., & Sniezek, J. A. (2003, May). The impact of information distribution, ownership, and discussion on group member judgment: The differential cue weighting model. *Organizational Behavior & Human Decision Processes, 91*(1), 12–26.

Chu, S. (2001, February). The possibilities are wireless: Designing and delivering information in the wireless space. *Technical Communication, 48*(1), 49–58.

Civin, M. A. (2000). *Male, female, email: The struggle for relatedness in a paranoid society.* New York: Other Press.

Clyman, J., Muchmore, M. W., & Sarrel, M. D. (2003, March 25). A host of options. *PC Magazine, 22*(5). Accessed Academic Search Elite.

Cohen, A. (2001, November 12). When terror hides online. *Time, 158*(21), 65.

Cole, J. 2000. *Surveying the digital future.* Los Angeles: UCLA Center for Telecommunications Policy (www.ccp.ucla.edu)

Coleman, L. H., Paternite, C. E., & Sherman, R. C. (1999, January). A reexamination of deindividuation in synchronous computer-mediated communication. *Computers in Human Behavior, 15*(1), 51–65.

Colombo, G. (2000, June). Polish your e-mail etiquette. *Sales and Marketing Management, 152*(6), p. 34.

Comber, C., Colley, A., Hargreaves, D. J., & Dorn, L. (1997). The effects of age, gender and computer experience upon computer attitudes. *Educational Research, 39,* 123–133.

Comeaux, P. (Ed.). (2002). *Communication and collaboration in the online classroom: Examples and applications.* Bolton, MA: Anker Publishing.

Compaine, B. (2001). *Re-examining the digital divide.* Cambridge, MA: MIT Press.

Conhaim, W. W. (2001, July–August). The Internet. *Link-up, 18*(4), 3, 8. *Constructivism, Instructivism, and Related Sites.* [On-line] Available: *http://www.emtech.net/construc.htm*

Cooper, A. & Sportolari, L. (1997). Romance in cyberspace: Understanding online attraction. *Journal of Sex Education and Therapy, 22*(1), 7–14.

Cowan, J. (2003). *Educating Americans about guns and safety.* Americans for Gun Safety. Available at: http://w3.agsfoundation.com/ Accessed May 5, 2003.

Cox, B. (1997, August). Evolving a distributed learning community, the online classroom in K12. *LN Magazine* [On-line]. 1(2). Available: http://www.aln.org/alnweb/magazine/issue2/cox.htm

Crockett, R. O. (2001, November 12). It's a phone, it's a handheld, it's . . . *Business Week, 3757,* 134.

Crystal, D. (2001). *Language and the Internet.* Cambridge: Cambridge University Press.

Csikszentmihalyi, M., & Kubey, R. W. (1990). *Television and the quality of life: How viewing shapes everyday experience.* Mahwah, NJ: Lawrence Erlbaum Associates.

Cyber play the taxpayer way: An analysis of Internet policy. (2000, Spring). Options for state government. *Spectrum: The Journal of State Government, 73*(2), 22.

Czitrom, D. (1982). *Media and American Mind: From Morse to McLuhan.* Chapel Hill: University of North Carolina Press.

Dahlgren, P. (2000, October–December). The Internet and the democratization of civic culture. *Political Communication, 17*(4). Accessed via Academic Search Elite.

Dale, P. (1976). *Language development: Structure and function.* New York: Holt, Rinehart & Winston.

Davidow, W. H., & Malone, M. S. (1992). *The virtual corporation.* New York: HarperCollins.

Davidson-Shivers, G., & Tanner, E. (2000). *Online Discussion: How Do Students Participate?* (ERIC Document Reproduction Service No. ED 443 410)

De Palma, P. (2001, June). Why women avoid computer science. *Association for Computing Machinery. Communications of the ACM, 44*(6), 27–29.

Deal, T. E., & Kennedy, A. A. (1982). *Corporate cultures: The rites and rituals of corporate life.* Reading, MA: Addison-Wesley.

DeKerckhove, D. (1997). *Connected intelligence: The arrival of the Web society.* Toronto: Somerville House.

Delacote, G. (1995, December/1996, January). The new frontiers of learning. *UNESCO Sources, 75,* 14.

Dery, M. (1994). (Ed.). *Flame wars: The discourse of cyberculture.* Durham, NC: Duke University Press.

Dickinson, K. (2003). *Cyberpl****y: Communication online.* Oxford, NY: Berg Publications.

DiMaggio, P. (2001). *Social implications of the Internet.* Paper Presented at Telecommunication Policy Res. Conference Arlington, VA.

DiMaggio, P., Hargittai, E., Neuman, W. R., & Robinson, J. P. (2001). Social implications of the Internet. *Annual Review of Sociology, 27*(1), 307–338.

Doherty, C., & Mayer, D. (2003, April). E-mail as a "contact zone" for teacher-student relationships. *Journal of Adolescent & Adult Literacy, 46*(7), 592–601.

Donath, J. S. (1999). Identity and deception in the virtual community. In P. Kollock & M. Smith (Eds.), *Communities in cyberspace.* London: Routledge.

Downloading hate. (1999, November 13). *The Economist, 353*(8145), 30–31.

Doyle, T., & Gotthoffer, D. (2000). *Quick guide to the Internet for speech communication* (2000 edition). Boston: Allyn & Bacon.

Durso, F. T., Hackworth, C. A., Barile, A. L., Dougherty, M. R. P., & Ohrt, D. D. (1998, Spring). Source monitoring in face-to-face and computer-mediated environments. *International Journal of Cognitive Technology, 3*(1), 32–38.

Dutton, W. H. (1999). *Society on the line: Information politics in the digital age.* New York: Oxford University.

Dyck, J. L., Gee, N. R., & Smither, J. A. (1998, January). The changing construct of computer anxiety for younger and older adults. *Computers in Human Behavior, 14*(1), 61–77.

Dyrli, O. E. (2000, May). How to best use the multilingual web. *Curriculum Administrator, 36*(5), p. 25.

Eagly, A. H. (1987). *Sex differences in social behavior: A social-role interpretation.* Hillsdale, NJ: Lawrence Erlbaum Associates.

Education Act. (1990). Revised Regulations of Ontario. Regulation 298. Available at: http://192.75.156.68:81/ISYSquery/IRLB9C8.tmp/4/doc

Elkin, T. (2001). Video-game marketing: $400 mil. *Advertising Age, 72*(40), 3–5.

Ess, C. (2001). Introduction: What's culture got to do with it? Cultural collisions in the electronic global village, creative interferences, and the rise of culturally-mediated computing. In C. Ess & F. Sudweeks (Eds.), *Culture, technology, communication: Towards an intercultural global village.* Albany, NY: State University of New York Press.

Estabrook, N. (1997). *Teach yourself the Internet in 24 hours.* Indianapolis, IN: Sams.Net.

Falling through the net: Toward digital inclusion. (2000). NTIA (National Telecommunications and Information Administration). Washington, DC: US Department of Commerce. Available at: http://www.ntia.doc.gov/ntiahome/fttn00/contents00.html Accessed May 6, 2003.

Fayol, H. (1937). The administrative theory in the state. Gree Trans. In L. Gulick & L. Urwick (Eds.), *Papers on the science of administration.* New York: Institute of Public Administration.

Feldman, A., Konold, C., & Coulter, B. (1999). Network science: A decade later. *Hands On! 22*(2), 1.

Fernback, J. (1997). The individual within the collective: Virtual ideology and the realization of collective principles. In S. Jones (Ed.), *Virtual culture: Identity and communication in cybersociety.* Thousand Oaks, CA: Sage Publications.

Ferrigno-Stack, J., Robinson, J. P., Kestnbaum, M., Neustadtl, A., & Alvarez, A. (2003, Spring). Internet and society. *Social Science Computer Review, 21*(1), 73–118.

Festinger, L. (1957). *A theory of cognitive dissonance.* Stanford, CA: Stanford University Press.

Fidler, R. (1997). *MediaMorphosis: Understanding NewMedia.* Thousand Oaks, CA: Pine Forge Press.

Fiedler, F. (1967). *A theory of leadership effectiveness.* New York: McGraw-Hill.

Fight hoaxters with the urban legend combat kit. (2000, November). *Inside the Internet, 7*(11), 14.

Filkins, D. (2001, September 25). As thick as the ash, myths are swirling. *New York Times, B,* B.8.

Flanders, V., & Willis, M. (1996). *Web pages that suck: Learn good design by looking at bad design.* San Francisco: Sybex.

Fletcher-Flinn, C. M., & Suddendorf, T. (1996a). Computer attitudes, gender and exploratory behavior: A developmental study. *Journal of Educational Computing Research, 15,* 369–392.

Fletcher-Flinn, C. M., & Suddendorf, T. (1996b). Do computers affect 'the mind'? *Journal of Educational Computing Research, 15*(2), 97–112.

Follett, M. (1986). Management as a profession. In M. T. Matteson & J. M. Ivancevich (Eds.), *Management and organizational behavior classics.* (4th ed.) Homewood, IL: BPI-Irwin, (pp. 8–9).

Fouser, R. J. (2001). "Culture," computer literacy, and the media in creating public attitudes toward CMC in Japan and Korea. In C. Ess & F. Sudweeks (Eds.), *Culture, technology, communication: Towards an intercultural global village.* Albany, NY: State University of New York Press.

Fox, S. A. (2000). The uses and abuses of computer-mediated communication for people with disabilities. In D. O. Braithwaite & T. L. Thompson (Eds.), *Handbook of communication and people with disabilities: Research and application.* (319–336). Mahwah, NJ: Lawrence Erlbaum Associates.

Freed, L. (1999, October 19). Internet access for all. *PC Magazine, 18*(18), 73.

Frey, L., Botan, C., Friedman, P., & Kreps, G. (1992). *Interpreting communication research: A case study approach.* Englewood Cliffs, NJ: Prentice Hall.

Funk, J. B., & Buchman, D. D. (1996). Playing violent video and computer games and adolescent self-concept. *Journal of Communication, 46*(2),19.

Gackenbach, J. (Ed.). (1999). *Psychology and the Internet: Intrapersonal, interpersonal, and transpersonal implications.* Oxford: Oxford University Press.

Gale, C. (2000, January). Online learning: A student perspective. *Syllabus, 13,* 52–53.

Ganesh, S. (2000, January). Mediating the imagination: Corporate involvement in the production of centralized subjectivity. *Journal of Communication Inquiry, 24*(1), 67–87.

Gastil, J. (1990). Generic pronouns and sexist language: The oxymoronic character of masculine generics. *Sex Roles, 23,* 629–643.

Gaud, W. S. (1999, November/December). Assessing the impact of web courses. *Syllabus, 13,* (4), 49–50.

Geller, L. (1999, March). The dawn of a new age in communications. *The Computer Post, 9*(3) [Online]. Available: http://www.cpost.mb.ca/march99/worldint/dawnage.htm

Gelman, R. B., & McCandlish, S. (1998). Electronic Frontier Foundation. In E. Gelman & E. Dyson (Eds.), *Protecting yourself online: The definitive resource on safety, freedom, and privacy in cyberspace.* New York: HarperCollins.

Gergen, K. J. (2000, Winter). The self in the age of information. *Washington Quarterly, 23*(1), 201–215.

Gilster, P. (1994). *The Internet navigator* (2nd ed.). New York: Wiley.

Gilster, P. (1997). *Digital literacy.* New York: Wiley.

Goffman, E. (1959). *Presentation of self in everyday life.* New York: Doubleday/Anchor Books.

Gold, L., & Maitland, C. (1999, April). *A review of contemporary research on the effectiveness of distance education in Higher Education.* Host name: Washington, DC: The Institute for Higher Education Policy. [On-line]. Available: http://www.ihep.com/PUB.htm.

Goldhamer, H. (1974). the social effects of communication technology. In W. Schramm & D. F. Roberts (Eds.), *The process and effects of mass communication.* Urbana: University of Illinois Press.

Golson, B. (2001, November 5). A bubble of bad news hides true Web reality. *Advertising Age, 72*(45)14.

Gozzi, R. (1999). *The power of metaphor in the age of electronic media.* Creskkill, NJ: Hampton.

Grandgenett, D. (2001, November). Problem resolution through electronic mail: A five-step model. *Innovations in Education & Teaching International, 38*(4), 347–354.

Greenhalgh, T. (2000, February 11). Artist weaves a wordy web. *Times Higher Educational Supplement, 1422,* p. 12.

Grice, H. P. (1975). Logic and conversation. In P. Cole & J. L. Morgan (Eds.), *Syntax and semantics: Vol. 3. Speech Acts* (pp. 113–128). New York: Academic Press.

Griffin, E. (2003). *A first look at communication theory* (5th ed.). Boston: McGraw-Hill.

Grimmes, B. (2001, December). Have web, don't travel. *PC World, 19*(12), 30–32.

Guernsey, L. (2000, August 9). Choice of Jewish candidate is noted in slurs on Internet, *New York Times, A,* p. A.18.

Gumpertz, J. J. (Ed.). (1982). *Language and social identity.* Cambridge, UK: Cambridge University Press.

Gurak, L. J. (2001). *Cyberliteracy: Navigating the Internet with awareness.* New Haven, CT: Yale University Press.

Hacker, K., & Steiner, R. (2001, Fall). Hurdles of access and benefits of usage for Internet communication. *Communication Research Reports, 18*(4), 399–407.

Hamman, R. B. (1999). Computer networks linking communities: A study of the effects of computer network use upon pre-existing communities. In U. Thiedke (Ed.), Virtualle Gruppen-Characteristika and Problemdimensionen (*Virtual groups: Characteristics and problematic dimensions*). Wiesbaden, Germany: Westdeutscher Verlag.

Hanks, W. F. (1996). *Language and communicative practices.* Boulder, CO: Westview.

Hanson, R., & Jubeck, T. (1999). Assessing the effectiveness of web page support in a large lecture course. *DEOSNEWS 9*(9) [On-line]. Available: http://www.ed.psu.edu/acsde/deos/deosnews/deosnews9_9.asp

Harmon, A. (2001, February 5). A new trick gives snoops easy access to e-mail. *New York Times, 150*(51655), p. C1.

Harmon, A. (2001, October 12). F.B.I. debunks e-mail threat. *New York Times. B,* B12.

Harmon, A. (2001, September 23). The search for intelligent life on the Internet. *New York Times, 4,* p. 4.2.

Harnad, S. (1995). Interactive cognition: exploring the potential of electronic quote/commenting. In B. Gorayska & J. L. Mey (Eds.), *Cognitive technology: in search of a humane interface.* Retrieved November 14, 2002 at http://www.cogsci.soton.ac.uk/~harnad/Papers/Harnad/Harnad95.interactive.cognition.html

Harris, J. (1994). *Way of the ferret: Finding educational resources on the Internet.* Eugene, OR: International Society for Technology in Education.

Harrison, S. A. (2001). The Internet, cyberadvocacy, and citizen communication. *Vital Speeches, 67*(20), 624.

Hawisher, G. E., & Selfe, C. L. (Eds.). (1999). *Passions, pedagogies, and 21st century technologies.* Logan, UT: Utah State University Press.

Haythornthwaite, C., Wellman, B., & Mantei, M. (1995). Work relationships and media use: A social network analysis. Special issue: Distributed communication systems. *Group Decision & Negotiation, 4*(3) 193–211.

Head, S. (1976). *Broadcasting in America: A survey of television and radio* (3rd ed.). Boston: Houghton Mifflin.

Heider, F. (1958). *The psychology of interpersonal relations.* New York: Wiley.

Heim, M. (1993). *The metaphysics of virtual reality.* Oxford: Oxford University Press.

Heim, M. (1999). The essence of VR. In V. Vitanza (Ed.), *CyberReader* (2nd ed.). Boston: Allyn and Bacon.

Henry, P. D. (2002, Fall). Scholarly use of the Internet by faculty members: Factors and outcomes of change. *Journal of Research on Technology in Education, 35*(1), 49–58.

Herbert, B. (2001, August 20). High-decibel hate. *New York Times, A,* p. A.17.

Herring, S. (1993). Gender and democracy in computer-mediated communication. *Electronic Journal of Communication 3*(2), 1–12.

Herring, S. (1999). Interactional coherence in CMC. *Journal of Computer-Mediated Communication, 4*(4).

Herring, S. (2001). Foreword. In C. Ess & F. Sudweeks (Eds.), *Culture, technology, communication: Towards an intercultural global village.* Albany: State University of New York Press.

Hewitt, B., Biddle, N., Simmons, M., & Tharp, M. (2002, August 26). Web of horror. *People, 58*(9), 67–69.

Hill, P., Lake, R., Celio, M., Campbell, C., Herdman, P., & Bulkley, C. (2001). *A study of charter school accountability.* U.S. Department of Education. Available online at: http://www.ed.gov/pubs/chartacct/chartacct_toc_summ.pdf

Hiltz, S., & Turoff, M. (1978). *The network nation: Human communication via computer.* Reading, MA: Addison-Wesley.

Hobman, E., Bordia, P., Irmer, B., & Chang, A. (2002, August). The expression of conflict in computer-mediated and face-to-face groups. *Small Group Research, 33*(4), 439–466.

Hoffman, D. L., & Novak, T. P. (1998). Bridging the racial divide on the Internet. *Science, 280,* 390–391.

Holeton, R. (Ed.). (1998). *Composing cyberspace.* Boston: McGraw Hill.

Honeycutt, L. (2001, January). Comparing e-mail and synchronous conferencing in online peer response, *Written Communication, 18*(1), 26–61.

Horton, D., & Wohl, R. R. (1956). Mass communication and parasocial interaction: Observations on intimacy at a distance. *Psychiatry, 19*(3), 215–229.

Howard, L. A., & Geist, P. (1995). Ideological positioning in organizational change: The dialectic of control in merging organization. *Communication Monographs, 62,* 110–131.

Howard, P. E. N., Rainie, L., & Jones, S. (2001, November). Days and nights on the internet: The impact of a diffusing technology. *American Behavioral Scientist, 45*(3), 383–405.

Huizinga, J. (1950). *Homo Ludens: A study of the play element in culture.* Boston: Beacon.

Infante, D. A., Rancer, A. S., & Womack, D. F. (1996). *Building communication theory* (3rd ed.). Prospect Heights, IL: Waveland Press.

Ingber, D. (1981, July). Computer addicts. *Science Digest, 89*–114.

Innis, H. A. (1951). *The bias of communication.* Toronto: University of Toronto Press.

Introna, L. D., & Nissenbaum, H. (2000). Shaping the Web: Why the politics of search engines matters. *The Information Society, 16*(3), 169–185.

Jackson, L. A. (1999, August). *Who's on the Internet: Making sense of Internet demographic surveys.* Paper presented at the symposium titled Conducting research on the Internet. American Psychological Association Convention, Boston, MA.

Jackson, L. A. (1999, October 14–16). *Social psychology and the digital divide.* Paper presented at the symposium titled "The Internet: A place for social psychology." The 1999 Conference of the Society of Experimental Social Psychology, St. Louis, MO.

Jackson, L. A., Ervin K. S., Gardner, P. D., & Schmitt, N. (2001, March). Gender and the Internet: Women communicating and men searching. *Sex Roles: A Journal of Research,* 363. Accessed via Expanded Academic ASAP.

Janis, I. (1982). *Victims of groupthink: A psychological study of foreign decisions and fiascos.* Boston: Houghton Mifflin.

Janney, A. W. (2000, Summer). Oedipus e-mails his mom: Computer-mediated romance develops as a science. *Extrapolation, 41*(2), 161–174

Janssen Reinen, I. A. M., & Plomp, T. J. (1997). Information technology and gender equality: A contradiction in terminus. *Computers & Education, 28,* 65–78.

Jamieson, P., Fisher, K., Gilding, T., Taylor, P. G., Trevitt, A. C. F. (2000, July). Place and space in the design of new learning environments. *Higher Education Research & Development, 19*(2). Accessed via Academic Search Elite.

Jenkins, H. (2001, January). The kids are all right online. *Technology Review, 104*(1), 121.

Johnscher, C. (1999). *The evolution of wired life: From the alphabet to the soul-catcher chip—how information technologies change our world.* New York: Wiley.

Johnson, K. (2000, August 9). Anti-Semitic hate groups take aim at Lieberman, *USA Today,* p. A.4.

Joinson, A. N. (2001, March-April). Self-disclosure in computer-mediated communication: The role of self-awareness and visual anonymity. *European Journal of Social Psychology, 31*(2), 177–192.

Jones, S. G. (Ed.). (1995). *CyberSociety: Computer-mediated communication and community.* Thousand Oaks, CA: Sage.

Jones, S. (1997a). The Internet and its social landscape. In S. Jones (1997) (Ed.), *Virtual culture: Identity and communication in cybersociety.* Thousand Oaks, CA: Sage Publications.

Jones, S. (Ed.). (1997b). *Virtual culture: Identity and communication in cybersociety.* Thousand Oaks, CA: Sage Publications.

Jones, S. (Ed.). (1999). *Doing Internet research: Critical issues and methods for examining the net.* Thousand Oaks, CA: Sage Publications.

Jung, J., Qiu, J. L., & Kim, Y. (2001, August). Internet connectedness and inequality: Beyond the "divide." *Communication Research, 28*(4), 507–536.

Kakutani, M. (2001, October 20). Fear, the new virus of a connected era. *New York Times, A,* p. A.13.

Kalmbach, J. (1997). *The computer and the page.* Norwood, NJ: Ablex.

Kaminer, W. (2001, November 25). Virtual rape. *New York Times Magazine, 151*(51948), 70–74.

Kasket, E. (2003, January). Online counselling. *Journal of the Society for Existential Analysis, 14*(1), 60–75.

Katz, D., & Kahn, R. L. (1978). *The social psychology of organizations* (2nd ed.). New York: Wiley.

Katz, J. E., & Aspden, P. (1997). A nation of strangers? *Communications of the ACM, 40,* 81–86.

Kelly, K. (1998). *New rules for the new economy: 10 radical strategies for a connected world.* New York: Penguin Viking.

Kelly, L., Duran, R. L., & Zolten. J. J. (2001). The effect of reticence on college students' use of electronic mail to communicate with faculty. *Communication Education, 50*(2), 170–176.

Kendall, N. M. (1998, March 16). Seeing MUD's gloppy promise. *Christian Science Monitor, 90*(75), 9.

Kent, M. L. (2001). Managerial rhetoric as the metaphor for the World Wide Web. *Critical Studies in Media Communication, 18*(3), 359–375.

Khan, B. H. (2000, February). Discussion of resources and attributes of the Web for the creation of meaningful learning environments. *CyberPsychology & Behavior, 3*(1), 17–24.

Kilian, C. (1997). F2F—Why teach online. *Educom Review 32*(4). [Online], Available: *http://www.educause.edu/pub/er/review/reviewArticles/32431.html*

Klaw, E., Dearmin Huebsch, P., & Humphreys, K. (2000, September). Community patterns in an on-line mutual help group for problem drinkers. *Journal of Community Psychology, 28*(5), 535–546.

Kohut, A. (2000, January/February). Internet users are on the rise; but public affairs interest isn't. *Columbia Journalism Review, 38*(5), 68–70.

Kopper, G., Kolthoff, A., & Czepek, A. (2000). Research review: Online journalism—A report on current and continuing research and major questions in the international discussion. *Journalism Studies, 1*(3), 499–512.

Kornblum, J. (2000, August 21). Cat lovers, felines get their paws on e-mail. *USA Today,* Life section, D.3.

Kramarski, B., & Feldman, Y. (2000, September). Internet in the classroom: Effects on reading comprehension, motivation and metacognitive awareness. *Educational Media International, 37*(3). Accessed Academic Search Elite.

Kraut, R., Patterson, M., Lundmark, V., Kiesler, S., Mukopadhyay, T., & Scherlis, W. (1998). Internet paradox: A social technology that reduces social involvement and psychological well-being? *American Psychologist, 53,* 1017–1031.

Kraut, R. E., Mukhopadhyay, T., Szczypula, J., Kiesler, S., & Schleris, B. (2000). Information and communication: Alternative uses of the Internet in households. *Information Systems Research, 10,* 287–303.

Kraut, R., Kresler, S., Boneva, B., Cummings, J., Helgeson, V., & Crawford, A. (2002, January). Internet paradox revisited. *Journal of Social Issues, 58*(1), 49–75.

Kreuter, M., Farrell, D., Olevitch, L., & Brennan, L. (2000). *Tailoring health messages: Customizing communication with computer technology.* Mahwah, NJ: Lawrence Erlbaum Associates.

Kuehn, S. A. (1993). Communication innovation on a BBS: A content analysis. *IPCT: International Computing and Technology, an Electronic Journal for the 21st Century.* Available as electronic ms. from IPCT at listserv@guvm.ccf.georgetown.edu. Message: get kuehn ipctv1n2.

Kunde, D. (1998, September 13). Colleges adding training to prepare web experts. *Lexington Herald-Leader,* E5.

Kushner, D. (2000, April). Untangling the Web's languages. *New York Times, 149*(51364), p. G1.

Kwok-Chi, N. (2001, June). Using e-mail to foster collaboration in distance education. *Open Learning, 16*(2). Accessed Academic Search Elite.

Lakoff, G., & Johnson, M. (1980). *Metaphors we live by.* Chicago, IL: University of Chicago Press.

Lambert, S., & Howe, W. (1993). *Internet basics: Your online access to the global electronic superhighway.* New York: Random House.

Lambert, B. (2001, November 22). Woman pleads guilty in abuse of teenager lured on Internet. *New York Times, 151*(51945), D5.

Lambert, B. (2003, February 12). 150-year sentence in sex abuse of teenager after Internet meeting. *New York Times, 152*(52392), B7.

Lamude, K. G., & Larsen, R. (1998, December). E-mail influence messages of Type-A scoring managers. *Perceptual & Motor Skills, 87*(3), 1246–1249.

Lane, D., & Shelton, M. (2001, July). The centrality of communication education in classroom computer-mediated communication: Toward a practical and evaluative pedagogy. *Communication Education, 50*(3), 241–255.

Langer, S. K. (1962). *Philosophy in a new key: A study in the symbolism of reason, rite, and art.* New York: Mentor Books.

Lanham, R. (1993). *The electronic word: Democracy, technology, and the arts.* Chicago: University of Chicago Press.

LaRose, R., Eastin, M. S., & Gregg, J. (2001). Reformulating the Internet paradox: Social cognitive explanations of Internet use and depression. *Journal of Online Behavior, 1*(2). Retrieved from the World Wide Web: http://www.behavior.net/job/v1n2/paradox.html

Lave, J., & Wenger, E. (1991). *Situated learning: Legitimate peripheral participation.* New York: Cambridge University Press.

Lawrence, P. R., & Lorsch, J. W. (1967). *Organization and environment: Managing differentiation and integration.* Boston: Harvard University.

Lea, M., O'Shea, T., Fung, P., & Spears, R. (1992). "Flaming" in computer-mediated communication: Observations, explanations, and implications. In M. Lea (Ed.) *Contexts of computer-mediated communication* (pp. 89–112). London: Harvester Wheatshaft.

Lea, M., & Spears, R. (1995). Love at first byte? Building personal relationships over computer networks. Under-studied relationships: Off the beaten track. In J. T. Wood & S. Duck (Eds.), *Understanding relationship processes series, 6* (pp. 197–233). Thousand Oaks, CA: Sage Publications.

Lee, J. Y. (1996). Charting the codes of cyberspace: A rhetoric of electronic mail. In L. Strate, R. Jaconsom, & S. Gibson (Eds.), *Communication and cyberspace* (pp. 275–296). Cresskill, NJ: Hampton.

Lewin, A., Lippitt, R., & White, W. (1939). Patterns of aggressive behavior in experimentally created social climates. *Journal of Social Psychology, 10,* 271–299.

Lewis, L., Snow, K., Farris, E., Levin, D., & Greene, B. (1999). *Distance education at postsecondary education institutions: 1997–98.* Washington, DC: National Center for Education Statistics Institute of Education Sciences. [Online] Available: http://nces.ed.gov/pubsearch/pubsinfo.asp?pubid=2000013.

Lieb, T. (1999, November/December). Incorporating multimedia into the writing course. *Syllabus, 13,*(4), 51–53.

Likert, R. (1967). *The human organization: Its management and value.* New York: McGraw-Hill.

Littlejohn, S. (2002). *Theories of human communication.* Belmont, CA: Wadsworth.

Littlejohn, S. W. (1977). Symbolic interactionism as an approach to the study of human communication, *Quarterly Journal of Speech, 63,* 84–91.)

Lucas, S. E. (1993). *The art of public speaking.* New York: McGraw-Hill.

Lynch, P. & Horton, S. (1997). *Yale style manual.* Yale University. [On-Line] Available: http://info.med.yale.edu/caim/manual/

Lynch, P. D., Kent, R. J., & Srinivasan, S. S. (2001, May/June). The global Internet shopper: Evidence from shopping tasks in twelve countries. *Journal of Advertising Research,41*(3), 15–23.

Mabry, M., & Thomas, E. (1991, July 22). A crisis of leadership. *Newsweek, 118*(4), 18–20.

MacKay, D. G. (1980). Psychology, prescriptive grammar, and the pronoun problem. *American Psychologist, 35,* 444–449.

Maine Department of Education. *Basic Computing,* (n.d.). Available: http://janus.state.me.us/education/technology/mods/mlist.html

Man posts bail in cyberspace rape case. (1997, February 25). *New York Times, 146*(50714), B6.

Mandel, T., & Van der Leun, G. (1996). *Rules of the net: Online operating instructions for human beings.* New York: Hyperion.

Manusov, V. (1995). Intentionality attributions for naturally occurring behaviors in intimate relationships. In J. Aitken & L. Shedletsky (Eds.), *Intrapersonal communication processes.* Annandale, VA: Speech Communication Association and Hayden-McNeil.

Marriott, M. (2000, March 23). Out of the mouths of babes: Wirelessly. *New York Times, G,* G.1

Martin, G. (1997, May). Getting personal through impersonal means: Using electronic mail to gain insight into student teacher's perceptions. *Research and Reflection, 3*(1).

McCollum, K. (1998b, October 16). Now that computers are the rule, U. of Florida begins to adopt. *The Chronicle of Higher Education.* A27–28.

McCombs, M., & Bell, T. (1996). The agenda-setting role of mass communication. In M. Sullivan & D. Stacks (Eds.), *An integrated approach to communication theory and research* (p. 89–105). Hillsdale, NJ: Lawrence Erlbaum Associates.

McCormick, N. B., & McCormick, J. W. (1992, Winter). Computer friends and foes: Content of undergraduates' electronic mail. *Computers in Human Behavior, 8*(4) 379–405.

McCown, J., Fischer, D., Page, R., & Homant, B. (2001). Internet relationships: People who meet people. *CyberPyschology & Behavior, 4*(5), 593–596.

McDermott, I. E. (2001, January). E-mail for everyone: Free from your friendly librarian. *Searcher, 9*(1), 59–64.

McDonald, M. (1999). Cyberhate: Extending persuasive techniques of low credibility sources to the World Wide Web. In D. W. Schumann & E. Thorson (Eds.), *Advertising and the World Wide Web: Advertising and consumer psychology* (pp. 149–157). Mahwah, NJ: Lawrence Erlbaum Associates.

McGregor, D. (1960). *The human side of enterprise.* New York: McGraw-Hill.

McIntosh, D. (1995). *Self, person, world: The interplay of conscious and unconscious in human life.* South Bend: Northwestern University Press.

McKelvey, T. (2001, July 16). Father and son target kids in a confederacy of hate. *USA Today,* D. 3.

McKenna, K. Y. A., & Bargh, J. A. (1998). Coming out in the age of the Internet: Identity "demarginalization" through virtual group participation. *Journal of Personality & Social Psychology, 75*(3), 681–694.

McKenna, K. Y. A., & Bargh, J. A. (1998, September). Coming out in the age of the Internet: Identity "demarginalization" through virtual group participation. *Journal of Personality & Social Psychology, 75*(3), 681–694.

McLuhan, M. (1964). *Understanding media: The extensions of man.* New York: McGraw-Hill.

McMillan, S. J. (1999). Health communication and the Internet: Relations between interactive characteristics of the medium and site creators, content, and purpose. *Health Communication, 11*(4), 375–390.

McMurdo, G. (1995). Getting wired for McLuhan's cyberculture. *Journal of Information Science, 21*(5), 371–382.

McRae, K., & Cree, G. S. (1999, December). Further evidence for feature correlations in semantic memory. *Canadian Journal of Experimental Psychology, 53*(4), 360–374.

Medintz, S. (2000, June). Watch your tongue. *Money, 29*(6), p, 34.

Messmer, M. (2001, April). Managing while out of the office. *Strategic Finance, 82*(10), 8–10.

Miller, P. J., & Hoogstra, L. (1992). Language as tool in the socialization and apprehension of cultural meanings. In T. Schwartz, G. White, & C. Lutz (Eds.), *New directions in psychological anthropology* (pp. 83–101). Cambridge, UK: Cambridge University Press.

Miller, P., & Jones, R. (1999). Online option extends classroom. *Security Management, 43*(5), 45.

Miller, S., & Fredericks, M. (2000). Social science research findings and educational policy dilemmas: Some additional distinctions. *Education Policy Analysis Archives, 8,*(3) [Online]. Available: http://epaa.asu.edu/epaa/v8n3/

Moe, W. W., & Fader, P. S. (2001, Summer). Uncovering patterns in cybershopping. *California Management Review, 43*(4), 106–117.

Mok, C. (1996). *Designing business: Multiple media, multiple disciplines.* San Jose: Adobe Press.

Montgomery, B., & Baxter, L. (1998). *Dialectical approaches to studying personal relationships.* Mahwah, NJ: Lawrence Erlbaum.

Morkes, J., Kernal, H. K., & Nass, C. (1999). Effects of humor in task-oriented human-computer interactions and computer-mediated communication: A direct test of SRCT theory. *Human-Computer Interaction, 14*(4), 395–435.

Morrell, R. W., & Echt, K. V. (1997). Designing written instructions for older adults: Learning to use computers. In A. D. Fisk & W. A. Rogers (Eds), *Handbook of human factors and the older adult.* (pp. 335–361). San Diego, CA: Academic Press, p. 419.

Mowlana, H. (1995). The communications paradox. *Bulletin of the Atomic Scientists, 51*(4), 40.

Murray, D. E. (1991). *Conversation for action: The computer terminal as a medium of communication.* Philadelphia: John Benjamins Publishing.

Neher, W. W. (1997). *Organizational communication: Challenges of change, diversity, and continuity.* Boston: Allyn and Bacon.

Nelson, L. J., & Cooper, J. (1997, May). Gender differences in children's reactions to success and failure with computers. *Computers in Human Behavior, 13*(2), 247–267.

Nesson, C. (2001, Spring). Threats to privacy. *Social Research, 68*(1), 105–113.

Netchaeva, I. (2002, October). E-government and e-democracy. *International Journal for Communication Studies, 64*(5), 467–478.

Newhagen, J., Cordes, J. W., & Levy, M. R. (1995, Summer). Nightly-nbc.com: Audience scope and the perception of interactivity in viewer mail on the Internet. *Journal of Communication, 45*(3) 164–175.

Newton, P., & Beck, E. (1993). Computing an ideal occupation for women? In J. Benyon & H. MacKay (Eds.), *Computers in classrooms: More questions than answers* (pp. 130–146). London: Falmer.

Nightline. (1999). An eight-week series of online discussions produced in conjunction with Forbes ASAP and ABCNEWS' Nightline in Primetime series Brave New World.

Nofsinger, R. E. (1999). *Everyday conversation.* Prospect Heights, IL: Waveland.

Nomura Research Institute. (1998). *Joho tsushin riyo ni kansuru kokusai hikaku chosa o jisse* ("Results of a comparative international survey on information technology usage"), 12 February [Online]. Available: http://www.nri.co.jp/news/1998/980212/index.html

Noonan, R. J. (1998). The psychology of sex: A mirror from the Internet. In Jayne Gackenbach (Ed.), *Psychology and the Internet: Intrapersonal, interpersonal, and transpersonal implications* (pp. 143–168). San Diego, CA: Academic Press.

Nussbaum, J., Pecchioni, L., Robinson, J., & Thompson, T. (2000). *Communication and aging* (2nd ed.). Mahwah, NJ: Lawrence Erlbaum Associates.

Olsen, F. (2001, June 8). Michigan deactivates internet program linked to several stalking incidents. *The Chronicle of Higher Education, 47*(39), A34.

O'Neil, J. (1996, November). On surfing and steering the Net. A conversation with Crawford Kilian. *Educational Leadership, 54*(3), 12–17.

Ong, W. J. (1982). *Orality and literacy: The technologizing of the word.* London: Methuen.

Oravec, J. A. (2002, January). Constructive approaches to Internet recreation in the workplace. *Communications of the ACM, 45*(1), 60–64.

Ortner, S. B. (1984). Theory in anthropology since the sixties. *Comparative Studies in Society and History, 26*(1), 126–166.

Ouchi, W. (1981). *Theory Z: How American businesses can meet the Japanese challenge.* Reading, MA: Addison-Wesley.

Pack, T., & Page, L. (2003, January). Creating community. *Information Today, 20*(1), 27–29.

Palmer, M. T. (1995). Interpersonal communication and virtual reality: Mediating interpersonal relationships. In F. Biocca & M. R. Levy (Eds.), *Communication in the age of virtual reality* (pp. 277–299). Hillsdale, NJ: Lawrence Erlbaum Associates.

Panero, J. C., Lane, D. M., & Napier, H. A. (1997). The Computer Use Scale: Four dimensions of how people use computers. *Journal of Educational Computing Research, 16*(4), 297–315.

Papert, S. (1981, March). Computers and computer cultures. *Creative Computing*, 82–92.

Papert, S. A., & Negroponte, N. (1996). *The connected family: Bridging the digital generation gap.* Atlanta, GA: Longstreet Press.

Parker, L. (1999, February 23). New avenues aiding hate group numbers. *USA Today*, 04A.

Parks, J. B., Roberton, M. A., & Safire, W. (2000). Development and Validation of an instrument to measure attitudes toward sexist/nonsexist language. *Sex Roles, 42*(5/6), 415–439.

Parks, M. R., & Floyd, K. (1996). Making friends in cyberspace. *Journal of Communication, 46*(1), 80–97.

Parks, M., & Roberts, L. (1998). Making MOOsic—The Development of Personal Relationships on line and a comparison to their off line counterparts. *Journal of Social & Personal Relationships, 15*(4), 517–537.

Pavlik, J. (1998). *New media technology: Cultural and commercial perspectives* (2nd ed.). Boston: Allyn and Bacon.

Pawlowski, D. R. (1998, Fall). Dialectical tensions in marital partners' accounts of their relationships. *Communication Quarterly, 46*(4), 396.

Peiris, D. R., Gregor, P, & Alm, N. (2000). The effects of simulating human conversational style in a computer-based interview. *Interacting with Computers, 12*, 635–650.

Penn, G. (2000). *Enrollment management for the 21st century: Delivering institutional goals, accountability and fiscal responsibility.* Office of Educational Research and Improvement (ED), Washington, DC. (ERIC Document Reproduction Service No. ED 432 939)

Perlstein, L. (1998, October 18). Essay assessor may make the grade. *Lexington Herald-Leader*, E3.

Perse, E. M., & Dunn, D. G. (1998, Fall). The utility of home computers and media use: Implications of multimedia and connectivity. *Journal of Broadcasting & Electronic Media, 42*(4), 435–456.

Perse, E., & Ferguson, D. (2000, Fall). The benefits and costs of Web surfing. *Communication Quarterly, 48*(4), 343–359.

Pesce, M. (2000). *The playful world: How technology is transforming our imagination.* New York: Ballantine Books.

Petersen, A. (1999, December 6). Lost in the maze: The web's explosive growth poses a challenge for users: How to make the most of it. *Wall Street Journal*, p. R6.

Peterson, J. (1994, May/June). Internet security. *Mother Jones, 19*(3),12.

Peterson, M. M. (2001). A tangled web. *National Journal, 33*(18),1314–1318.

Petty, R. E., & Cacioppo, J. T. (1986). *Communication and persuasion: Central and peripheral routes to attitude change.* New York: Springer-Verlag.

Petty, R. E., Ostrom, T. M., & Brock, T. C. (1981). *Cognitive responses in persuasion.* Hillsdale, NJ: Lawrence Erlbaum Associates.

Phipps, R., & Merisotis, J. (1999). *What's the difference? A review of contemporary research on the effectiveness of distance learning in higher education.* Washington, DC: The Institute for Higher Education Policy [Online]. Available: http://www.ihep.com/Home.php

Poisoning the Web: Hatred online. (1999, August). *Corrections Today, 61*(5), 102–104.

Poole, T., & Hansen, J. (2001, January). Tips for tracking the e-mail trail. *Security Management, 45*(1), p. 42–47.

Porter, P. (2001). The Internet as conference room. *Design News, 56*(8), para, 2–4.

Postman, N. (1993). *Technology: The surrender of culture to technology.* New York: Vintage.

Postman, N., & Paglia, C. (1999). She wants her TV! He wants his book! In V. Vitanza. (Ed.), *CyberReader* (2nd ed.). Boston: Allyn and Bacon.

Postman, N. (1986). *Amusing ourselves to death: Public discourse in the age of show business.* New York: Penguin.

Postmes, T., Spears, R., & Lea, M. (1998, December). Breaching or building social boundaries?: SIDE-effects of computer-mediated communications. *Communication Research, 25*(6), 689–715.

Postmes, T., Spears, R., & Lea, M. (2000). The formation of group norms in computer-mediated communication. *Human Communication Research, 26*(3), 341–72.

The power of speech. (2000, May 13). *Economist, 355*(8170), 60–2.

Power, R. (2000). *Tangled Web: Tales of digital crime from the shadows of cyberspace.* Bloomington: Indiana University.

Putnam, L. L., & Pacanowsky, M. E. (Eds.). (1983). *Communication and organizations: An interpretative approach.* Newbury Park: CA: Sage Publications.

Radford, M., Barnes, S., & Barr, L. (2002). *A student's guide for evaluating web sites.* Boston: Allyn and Bacon.

Rafaeli, S., & Sudweeks, F. (1996). Networked interactivity. *Journal of Computer-Mediated Communication, 2*(4).

Rawlins, W. (1992). *Friendship matters: Communication, dialectics, and the life course.* New York: Aldine de Gruyter.

Ray, D. S. (1999). Matching Internet resources to information needs. An approach to improving Internet search results. *Technical Communication, 46*(4), 569–574.

Redding, W. C. (1992). Stumbling toward identity: The emergence of organizational communication as a field of study. In K. L. Hutchinson (Ed.), *Readings in organizational communication* (pp. 11–44). Dubuque, IA: Wm. C. Brown.

Reeves, B., & Nass, C. (1999). *The media equation: How people treat computers, television, and new media like real people and places.* Stanford, CA: C S L I Publications.

Reid, E. M. (1991). *Electropolis: Communication and community on Internet relay chat.* Melbourne, Australia: University of Melbourne Department of History.

Reid, F. J., Malinek, V., Stott, C. J. T., & Evans, J. (1996, August). The messaging threshold in computer-mediated communication. *Ergonomics, 39*(8) 1017–1037.

Rheingold, H. (1993). *Virtual communities.* Reading, MA: Addison-Wesley.

Rheingold, H. (1997). *The virtual community* [Online]. Available: http://www.well.com/user/hlr/vcbook/

Rice, R. E., & Love, G. (1987). Electronic emotion: Socioemotional content in a computer-mediated network. *Communication Research, 14,* 85–108.

Rifkin, J. (1987). *Time wars.* New York: Touchstone Books.

Riphagen, J., & Kanfer, A. (1997). *How does email affect our lives?* Champaign-Urbana, IL: National Center for Supercomputing Applications, University of Illinois [Online]. Available: www.ncsa.uiuc.edu/edu/trg/e-mail/index.html

Robinson, J. P., & Kestnbaum, M. (1999, Summer). The personal computer, culture and other uses of free time. *Social Science Computer Review, 209–216.*

Robinson, J. P., Kestnbaum, M., Neustadtl, A., & Alvarez, A. (2000, May). IT, the Internet, and time displacement. Paper presented at the Annual Meeting of the American Association Public Opinion Research, Portland, OR.

Rocks, D., Pascual, A. M., Little, D., & Brown, J. (2001, October 29). The net as a lifeline. *Business Week, 3755,* pp. EB16–22.

Rogers, E. M. (1995). *Diffusion of innovations.* New York: Free Press.

Rogers, E. M., Daley, H. M., & Wu, T. D. (1983). *The diffusion of home computers.* Stanford, CA: Institute for Communication Research, Stanford University.

Rogers, M., & Oder, N. (2000). Librarian victim of Net "Slavemaster." *Library Journal, 125*(12), p. 26

Rose, L. (1995). *Netlaw: Your rights in the online world.* Berkeley, CA: Osborne McGraw-Hill.

Rovner, J. (1999). Doctors sue Internet anti-abortionists. *The Lancet, 353*(9149), 303.

Rubin, R. B., & McHugh, M. P. (1987). Develoment of parasocial interaction relationships. *Journal of Broadcasting and Electronic Media, 31,* 279–292.

Rubin, R. B., Perse, E, M, & Barbato, C. A. (1988, Summer). Conceptualization and measurement of interpersonal communication motives. *Human Communication Research, 14*(4), 602–628.

Russell, T. L. (2000). *The no significant difference phenomenon: A comparative research annotated bibliography on technology for distance education.* Raleigh, NC: North Carolina State University.

Rutkowski, A. (2000). *Understanding next generation Internet: An overview of developments.* Telcommunications Policy Online. Available at: http://www.tpeditor.com/contents/2000/rutkowski.htm Accessed May 6, 2003.

Saffo, P. (1992). Paul Saffo and the 30-year rule. *Design World, 24,* 18.

Safire, W. (1999, May 16). Genderese. *New York Times Magazine, 148*(51524), 30–32.

Sager, I., Carey, J., & Kerstetter, J. (2001, October 22). Preparing for a Cyber-Assault Fortification of the Web has become a top priority. *Business Week, 3754,* p. 50.

Sammons, M. (1999). *The Internet writer's handbook.* Boston: Allyn and Bacon.

Samoriski, J. (2002). *Issues in cyberspace.* Boston: Allyn and Bacon.

Sankis, L. M., & Widiger, T. A. (1999, December). Gender Bias in the English Language? *Journal of Personality & Social Psychology, 77*(6), 1289–1296.

Sapir, E. 1921. *Language.* New York: Harcourt, Brace & Co.

Sarbaugh-Thompson, M., & Feldman, M. S. (1998, November–December). Electronic mail and organizational communication: Does saying "hi" really matter? *Organization Science, 9*(6), 685–698.

Savicki, V., Kelley, M., & Oesterreich, E. (1999, March). Judgments of gender in computer-mediated communication. *Computers in Human Behavior, 15*(2), 185–194.

Sayers, J. (1986). *Sexual contradictions: Psychology, psychoanalysis, and feminism.* New York: Tavistock Publications.

Schaefer, D. R., & Dillman, D. A. (1998, Fall). Development of a standard e-mail methodology: Results of an experiment. *Public Opinion Quarterly, 62*(3), 378–397.

Schatz, B. R. (2002, January). The interspace: Concept navigation across distributed communities. *Computer, 35*(1), 54–63.

Schmidt, S. (1999). *Distance education 2010: A virtual space odyssey.* Paper prepared for the conference Teaching with Technology: Rethinking Traditions.

Schmidt, S., Shelley, M., Van Wart, M., Clayton, J., & Schleck, E. (2000). The challenges to distance education in an academic social science discipline: The case of political science. *Education Policy Analysis Archives, 8*(27) [Online]. Available: http://epaa.asu.edu/epaa/v8n27/

Schmitz, J., & Fulk, J. (1991). Organizational colleagues, media richness, and electronic mail. *Communication Research, 18,* 487–523.

Schofield, J. W. (1997). Psychology. Computers and classroom social processes: A review of the literature. *Social Science Computer Review, 15,* 27–39.

Schonsheck, J. (1997). Privacy and discrete "social spheres." *Ethics & Behavior, 7*(3), 1997, 221–228.

Schroeder, R. (1996). Playspace invaders: Huizinga, Baudrillard and video game violence. *Journal of Popular Culture, 30*(3), 143–153.

Schwartz, J. (2001, April 22). When online hearsay intrudes on real life. *New York Times, 3,* p. 3.10.

Science and technology: Picture perfect? (2001, October 20). Broadcasting technology. *The Economist, 361*(8244), 75–76.

Scodari, C. (1998, Fall). "No politics here": Age and gender in soap opera "cyberfandom." *Women's Studies in Communication, 21*(2), 168–187.

Shah, R., & Romine, J. (1995). *Playing MUDs on the Internet.* New York: Wiley.

Shannon, C., & Weaver, W. (1999). *The Mathematical Theory of Communication.* Chicago: University of Illinois Press.

Shedletsky, L. (1989). *Meaning and mind: An intrapersonal approach to human communication.* ERIC and The Speech Communication Association.

Shedletsky, L. (1993). Minding computer-mediated communication: CMC as experiential learning. *Educational Technology, 33*(12), 5–10.

Shedletsky, L. (1993, April). Computer-mediated communication to facilitate seminar participation and active thinking. *Electronic Journal of Communication/La Revue Electronique de Communication 3*(2), [Special Issue]. (CIOS members can find the paper at: EJC/REC. [On-Line] Available: http://www.cios.org/www/ejc/v3n293.htm

Shedletsky, L. J. (2000). The online double-bind: Mixed messages about online teaching and scholarship. *The Maine Scholar, 13,* 163–173.

Shedletsky, L. J., & Aitken, J. E. (2001, July). The paradoxes of online academic work. *Communication Education, 50*(3), 206–217.

Shedletsky, L. J., & Aitken, J. (2002). Intrapersonal communication, interpersonal communication, and computer-mediated communication: A synergetic collaboration. In P. Comeaux (Ed.), *Communication and collaboration in the online classroom: Examples and applications.* Bolton, MA: Anker Publishing.

Shedletsky, L., Ouillette, T., & Monfort, T. (2001, April). *Introducing senior citizens to the Internet and successful mentoring.* Paper presented at the Eastern Communication Association, panel on Instructional Practices, Portland, Maine.

Sheppard, R. (1999). Patrolling for hate on the Net. *Maclean's, 112*(2), 64.

Sherer, K. (1997, November–December). College life online: Healthy and unhealthy Internet use. *Journal of College Student Development, 38*(6), 655–665.

Sherry, L., Billig, S. H., & Tavalin, F. (2000). Good online conversation: Building on research to inform practice. *Journal of Interactive Learning Research, 11*(1), 85–127.

Shockley-Zalabak, P. (2002). Protean places: Teams across time and space. *Journal of Applied Communication Research, 30* (3), 231–250.

Sieber, U. (2001). Fighting hate on the Internet. Organisation for Economic Cooperation and Development. *The OECD Observer Paris, 224,* 64–66.

Siegel, D. (1996). *Creating killer web sites.* Indianapolis: Hayden Books, [Online]. Available: http://www.killersites.com.

Simmons, J. (2001, May). Facing today's challenges. *Risk Management, 48*(5), 21.

Skinner, B. F. (1987). Whatever happened to psychology as a science of behavior? *American Psychologist, 42,* 780–786.

Skinner, D. (2000, Winter). McLuhan's world—and ours. *Public Interest, 138,* 52–65.

Slater, J. (2002, November 22). Internet child sex hunters to be jailed. *Times Educational Supplement,* (4508), 11.

Slouka, M. (1995). *War of the worlds: Cyberspace and the high-tech assault on reality.* New York: Basic Books.

Smith, J. J., Keil, M., & Depledge, G. (2001, Fall). Keeping mum as the project goes under: Toward an explanatory model. *Journal of Management Information Systems, 18*(2), 189–227.

Soat, J. (2003, April 21). Privacy, security, identity still matter. *InformationWeek* (936), 75.

Sontag, S. (1966). *On culture and the new sensibility. Against interpretation.* New York: Dell Publishing.

Soukup, C. (1999). The gendered interactional patterns of computer-mediated chatrooms: A critical ethnographical study. *Information Society, 15*(3), 169–176.

Spinello, R. (2000). *Cyberethics: Morality and law in cyberspace.* Boston: Jones and Bartless.

Sproull, L., & Kiesler, S. (1986). Reducing social context cues: Electronic mail in organizational communication. *Management Science, 32,* 1492–1512.

Sproull, L., & Kiesler, S. (1991). *Connections: New ways of working in the networked organization.* Cambridge, MA: The MIT Press.

Stanek, W. (1998). *Learn the Internet in a weekend.* Rocklin, CA: Prima.

Staudenmeier, J. J., Jr. (1999, January). Children and computers. *Journal of the American Academy of Child & Adolescent Psychiatry, 38*(1), 5.

Stead, B. A., & Gilbert, J. (2001, November). Ethical issues in electronic commerce. *Journal of Business Ethics, 34*(2), 75–85.

Stefik, M. (1996). *Internet dreams: Archetypes, myths, and metaphors.* Cambridge, MA: MIT Press.

Stephenson, W. (1967). *The play theory of mass communication.* Chicago: University of Chicago Press.

Stephenson, W. (1988). *The play theory of mass communication.* New Brunswick, NJ: Transaction.

Stewart, C. M., Shields, S. F., & Sen, N. (2001). Diversity in on-line discussions: A study of cultural and gender differences in listservs. In C. Ess & F. Sudweeks (Eds.), *Culture, technology, communication: Towards an intercultural global village.* Albany, NY: State University of New York Press.

Stix, G. (2002, September). Real time. *Scientific American, 287*(3), 36–40.

Strand, E. A. (1999, March). Uncovering the role of gender stereotypes in speech perception. *Journal of Language & Social Psychology, 18*(1), 86–101.

Stross, R. E. (2001, November 11). The rumor mail. *U.S. News & World Report, 131*(20), 44.

Suler, J. (1996). *The psychology of cyberspace.* [Online]. Available: http://www.rider.edu/2sites/psycyber/psycyber.html

Suler, J. (2002). The online disinhibition effect. In *The psychology of cyberspace.* [Online]. Available: http://www.rider.edu/users/suler/psycyber/basicfeat.html

Sveningson, M. (2000). *Cyber love: Creating romantic relationships on the net.* Unpublished manuscript.

Swats, V. R., & Walters, T. L. (1995). *The U.S. mail vs. e-mail: Understanding interpersonal communication through traditional pedagogical pathways.* (ERIC Document Reproduction Service No. ED 393 127)

Sypher, B. D., Applegate, J. L., & Sypher, H. E. (1985). Culture and communication in organizational contexts. In W. G. Gudykunst, L. P. Stewart, & S. Ting-Toomey (Eds.), *Communication, culture, and organizational processes.* Beverly Hills, CA: Sage Publications.

Taguiri, R. (1968). The concept of organizational climate,. In R. Tagiuri & G. H. Bitwin (Eds.), *Organizational climate: Explorations of a concept.* Boston: Harvard University Press.

Tannen, D. (1990). *You just don't understand: Women and men in conversation.* New York: William Morrow.

Tao, L. (1995). *What do we know about email—An existing and emerging literacy vehicle?* (ERIC Document Reproduction Service No. ED 399 530)

Taylor, F. W. (1911). *The principles of scientific management.* New York: Harper & Brothers.

Taylor, D., & Altman, I. (1987). Communication in interpersonal relationships: Social penetration processes. In M. Roloff & G. Miller (Eds.), *Interpersonal processes: New directions in communications research.* Newbury Park, CA: Sage.

Teaching and learning for tomorrow: A learning technology plan for Maine's future. (2001, January). *Final report of the task force on the Maine learning technology endowment.* Available online at: http://www.state.me.us/governor/news/previous_articles/PressReleases/MLTE.pdf

Teich, A. H. (1997). *Technology and the future.* New York: St. Martin's Press.

Teske, J., (2002, September). Cyberpsychology, human relationships, and our virtual Interiors. *Zygon: Journal of Religion & Science, 37*(3), 677–701.

Thatcher, J. B., & Perrewe, P. L. (2002, December). An empirical examination of individual traits as antecedents to computer anxiety and computer self-efficacy. *MIS Quarterly, 26*(4), 381–397.

Thomas, J. (1999, August 16). New face of terror crimes: "Lone wolf" weaned on hate. *New York Times,* A. p. 1.

Thompsen, P. A. (1994, Spring). An episode of flaming: A creative narrative. *Etc: A Review of General Semantics, 51*(1), 51–72.

Thoreau, H. D. (1854; reprint 1951). Quoted in Roger Fidler (1997). *Mediamorphosis: Unersthanding new media.* Thousand Oaks, CA: Pine Forge Press.

Thorington, H. (1999, Summer). Loose ends/connections: Interactivity in networked space. *Style, 33*(2) 212–231.

Townsend, A. M. (2000, Summer). Solidarity.com? Class and collective action in the electronic village. Journal of Labor Research. *Fairfax, 21*(3), 393–405.

Turkle, S. (1984). *The second self: Computers and the human spirit.* New York: Simon & Schuster.

Turkle, S. (1995). *Life on the screen: Identity in the age of the Internet.* New York: Simon & Schuster.

Turkle, S. (1997). *Life on the screen: Identity in the age of the Internet.* New York: Touchstone.

Turkle, S. (1999). Identity crisis. In V. Vitanza (Ed.), *CyberReader* (2nd ed.). Boston: Allyn and Bacon.

Ullman, E. (1997). Close to the machine: *Technophilia and its discontents.* San Francisco: City Lights Books.

U.S. Department of Commerce, National Telecommunications and Information Administration (NTIA), Economics and Statistics Administration; U.S. Bureau of the Census Washington, DC. (2002, February). *A nation online: How Americans are expanding their use of the Internet.*

Uslaner E. (2001). *The Internet and social capital.* Proc. ACM. Forthcoming

Van Der Leun, G., & Mandel, T. (1996). *Rules of the Net: On-line operating instructions for human beings.* New York: Hyperion.

Van Dijk, J. (2000). Widening information gaps and policies of prevention. In K. Hacker & J. Van Dijk (Eds.), *Digital democracy: Issues of theory and practice* (pp. 166–183). London: Sage Publications.

Veryard, R. (1987, October). *Are computers sexy? Notes on technology and gender.* [Online]. Available: http://www.users.globalnet.co.uk/~rxv/tcm/gender.htm

Vitanza, V. (Ed.). (1999). *CyberReader* (2nd ed.). Boston: Allyn and Bacon.

VonBertalanffy, L. (1968). General theory of systems: Application to psychology. *Social Science Information, 6,* 125–136.

von Hippel, W., Sekaquaptewa, D., & Vargas, P. (1995). On the role of encoding processes in stereotype maintenance. *Advances in Experimental Social Psychology, 27,* 177–254.

Vygotsky, L. (1962). *Thought and language.* Translated by E. Hanfmann and G. Vakar. Cambridge, MA: MIT Press.

Vygotsky, L. S. (1982) Sobranie sochinenii, Tom pervyi: Voprosy teorii i istorii psikhologii [*Collected works, vol. I: Problems in the theory and history of psychology*]. Moscow: Izdatel'stvo Pedagogika.

Wajcman, J. (1991). Technology as masculine culture. In J. Wajcman (Ed.), *Feminism confronts technology.* State College: Penn State University Press.

Waldeck, J., Kearney, P. & Plax, P. (2001). Teacher e-mail message strategies and students' willingness to communicate. *Journal of Applied Communication Research, 12*(1), 54–70.

Walsh, N. E. (1981). *Understanding computers: What managers and users need to know.* New York: Wiley.

Walston, J. T., & Lissitz, R. W. (2000, October). Computer-mediated focus groups. *Evaluation Review, 24*(5), 457–483.

Walther, J. B. (1993, Fall). Impression development in computer-mediated interaction. *Western Journal of Communication, 57*(4), 381–398.

Walther, J. B. (1996, February). Computer-mediated communication: Impersonal, interpersonal, and hyperpersonal interaction. *Communication Research, 23*(1) 3–43.

Walther, J. B., Anderson, J. F., & Park, D. W. (1994). Interpersonal effects in computer-mediated interaction: A meta-analysis of social and antisocial communication. *Communication Research, 21*(4), 460–487.

Waterman, R. H. (1990). *Adhocracy: The power to change.* New York: W. W. Norton.

Watt, D., & White, J. M. (1999, Winter). Computers and the family life: A family development perspective. *Journal of Comparative Family Studies, 30*(1), 1–16.

Weber, M. (1947). *The theory of social and economic organization.* Henderson and Parsons, Trans. New York: The Free Press.

Wegner, S., Holloway, K., & Garton, E. (1999, November). The effects of Internet-based instruction on student learning. *ALN Magazine, 3* (2). [On-Line] Available: http://www.aln.org/alnweb/journal/Vol3_issue2/Wegner.htm

Weick, K. (1995). *Sensemaking in organizations.* Thousand Oaks, CA: Sage Publications.

Weiner, R. (2000). Instructors say online courses involve more work at same pay. *The New York Times on the*

Web, 6. Available at: http://www.nytimes.com/library/tech/00/06/cyber/education/21education.html Accessed May 6, 2003.

Weisband, S. P., Schneider, S. K., & Connolly, T. (1995, August). Computer-mediated communication and social information: Status salience and status differences. *Academy of Management Journal, 38*(4) 1124–1151.

Welch, K. E. (1999). *Electric rhetoric: Classical rhetoric, oralism, and a new literacy (digital communication).* Cambridge, MA: MIT Press.

Wellman, B., & Gulia, M. (1999). Virtual communities as communities: Net surfers don't ride alone. In M. A. Smith & P. Kollock (Eds.), *Communities in cyberspace.* New York: Routledge.

Wellner, A. S. (2001, October 1). Who isn't wired in the United States of America? *Forecast, 21*(16),11.

Werde, B. (2003, March 20). Is the employer real? Guarding your personal information. *New York Times, 152*(52428), E6.

Wheeler, D. (2001). New technologies, old culture: A look at women, gender, and the Internet in Kuwait. In C. Ess & F. Sudweeks (Eds.), *Culture, technology, communication: Towards an intercultural global village.* Albany, NY: State University of New York Press.

White, H., McConnell, E., Clipp, E., & Bynum. L. (1999, September). Surfing the net in later life: A review of the literature and pilot study of computer use and quality of life. *Journal of Applied Gerontology, 18*(3), 358–378.

Whitelaw, K. (1996, November 4). Fear and dread in cyberspace. *U.S. News & World Report, 121*(18) 50.

Whitley, B. E. (1997). Gender differences in computer-related attitudes and behaviour: A meta-analysis. *Computers in Human Behaviour, 13,* 1–22.

Whittle, D. B. (1997). *Cyberspace: The human dimension.* New York: W. H. Freeman.

Whitty, M., & Gavin, J. (2001). Age/sex/location: Uncovering the social cues in the development of online relationships. *CyberPsychology & Behavior, 2* (5), 623–630.

Whorf, B. L. (1941). The relation of habitual thought and behavior to language. In B. L. Whorf (Ed.), *Essays in memory of Edward Sapir.* Salt Lake City, Utah: University of Utah.

Whorf, B. L. (1956). *Language, thought, and reality.* Cambridge, MA: MIT Press.

Wiener, N. (1948). *Cybernetics: Control and communication in the animal and the machine.* Cambridge, MA: MIT Press.

Wildermuth, S. (2001). Love on the line: Participants' descriptions of computer-mediated close relationships. *Qualitative Research Reports in Communication, 2*(4), 89–95.

Wildstrom, S. H. (2001, November 12). A handheld for every pocket. *Business Week, 3757,* 124–127.

Wilhelm, A. B. (2000). *Democracy in the digital age.* New York: Routledge.

Winch, P. (1958). *The idea of a social science and its relation to philosophy.* Atlantic Highlands, NJ: Humanities Press.

Winograd, T., & Flores, F. (1987). *Understanding computers and cognitivion.* Reading, MA: Addison-Wesley.

Winston, B. (1995). How are media born and developed? In J. Downing, A. Mohammadi, & A. Sreberny-Mohammadi (Eds.), *Questioning the media: A critical introduction.*

Wolak, J., Mitchell, K. J., & Finkelhor, D., (2003, February). Escaping or connecting? Characteristics of youth who form close online relationships. *Journal of Adolescence, 26*(1), 105–120.

Wood, J. T. (1999). *Communication theories in action: An introduction* (2nd ed.). Belmont, CA: Wadsworth.

Woolley, B. (1999). Cyberspace. In V. Vitanza (Ed.). *CyberReader* (2nd ed.). Boston: Allyn and Bacon.

Wright, K. B., & Bell, S. B. (2003, January). Health-related support groups on the Internet: linking empirical findings to social support and computer-mediated communication theory. *Journal of Health Psychology, 8*(1), 39–55.

Wynn, E., & Katz, J. E. (1997, October–December). Hyperbole over cyberspace: Self-presentation and social boundaries in Internet home pages and discourse. *Information Society, 13*(4), 297–328.

Year 2000: How colleges are budgeting for technology. (2000, June 11). *News, Resources, and Trends* [Online] Available: http://www.syllabus.com/news/ntr_latest_news.cfm

Young, K. S. (1998). *Caught in the Net: How to recognize the signs of Internet addiction and a winning strategy for recovery.* New York: Wiley.

Zagorsky, J. L. (1997, Fall). E-mail, computer usage and college students: A case study. *Education, 118*(1), 47–56.

Zhifang, Z. (2002, May). Linguistic relativity and cultural communication. *Educational Philosophy & Theory, 34*(2), 161–171.

Zillmann, D. (1971). Excitation transfer in communication-mediated aggressive behavior, *Journal of Experimental Social Psychology, 7,* 419–434.

Author Index

A

Abbate, J., 225
Abdal-Haqq, I., 214
Ackermann, E., 52
Adkins, M.E., 128
Adler, R.B., 27, 76, 242
Administration on Aging, 135
Adrianson, L., 137, 203
Agre, P., 81, 182
Aitken, J.E., 24, 121, 125, 130, 153, 154, 170, 173, 213
Althaus, S.L., 182
Altman, I., 151
Andersen, P., 132
Anderson, J.F., 143, 151
Anderson, R.H., 96
Angeli, C., 201
Applegate, J.L., 188
Arakaki Game, J., 10, 78
Argyris, C., 191, 208
Armistead, T., 110
Arnold, A.M., 107
Arquette, T., 43
Arrow, H., 167
Artz, N., 111
Ashling, J., 202
Ashton, G., 52
Asman, M.F., 197
Aspden, P., 230
Aytes, K., 202

B

Bakhtin, M.M., 9
Barclay, D., 91
Bargh, J.A., 160, 172, 173, 239
Barksdale, K., 52
Barlow, J.P., 246
Barnard, C., 190
Barnatt, C., 191
Barnes, S.B., 19, 52, 53, 102, 106, 143
Baron, N., 41, 42, 150, 158, 229, 231
Barrett, E., 137
Barron, A., 52
Barron, M., 9, 25, 111
Basch, R., 54
Basso, E.B., 9
Bauman, R., 9
Baxter, L., 37, 38
Baym, N., 9, 76, 159, 162, 171, 172
Beck, E., 103
Beirne, M., 178
Bell, S.B., 178, 218, 219
Bell, T., 153
Belsare, T.J., 111, 244
Benokraitis, N.V., 105
Berlo, D.K., 123
Berman, D., 110
Berman, J., 100
Berry, W., 150, 151
Biddle, N., 157
Biggs, S., 156
Bikson, T.K., 96
Billig, S.H., 216
Bing, J., 103
Bishop, A., 53
Blackwell, R.D., 200
Blake, R., 197
Bodker, S., 167
Boggs, C., 105
Bohannan, P., 219
Bonebrake, K., 217
Bordia, P., 24, 100
Bormann, E.G., 81, 89, 168
Botan, C., 55
Bowen, C., 91, 108
Brandon, D.P., 173
Breen, R., 215

Brehm, J.W., 97, 98, 114
Brewer, S., 125
Brody, J.E., 91, 100
Bruening, P., 100
Bruner, M.S., 179
Brunyand, J.H., 100
Buchman, D.D., 98, 99
Bulkeley, D., 199
Bull, K.S., 154
Burke, K., 103, 202
Burwell, L.A., 205, 221
By the numbers, 205

C

Cacioppo, J.T., 131
California Department of Education, 209
Callahan, C., 62
Callaway Russo, T., 212
Camp, L.J., 182
Camp, T., 103, 167
Campbell, S., 123
Cardenas, K.H., 209
Carey, J., 44
Carroll, J.B., 80, 102
Cassell, M.M., 179
Cattagni, A., 213
Cavanagh, J., 191
Cavazos, E.A., 246
Cegala, D.J., 132, 168, 183
Chadwick, S.A., 212
Cherny, L.M., 80
Chernyshenko, O.S., 171
Cho, A., 110
Chu, S., 230
Civin, M.A., 133
Clyman, J., 12
Cohen, A., 110
Cole, J., 239
Coleman, L.H., 180
Colley, A., 137
Colombo, G., 241
Comber, C., 137
Comeaux, P., 218, 332
Compaine, B., 97
Conhaim, W.W., 172
Cooper, A., 151
Cooper, J., 138
Cordes, J.W., 147
Cowan, J., 98

Cox, B., 217, 219, 220
Cree, G.S., 132
Crockett, R.O., 168
Crystal, D., 102, 114
Csikszentmihalyi, M., 88
Cyber play the taxpayer way, 194
Czitrom, D., 91

D

Dale, P., 122
Daley, H.M., 231
Davidow, W.H., 199
Davidson-Shivers, G., 64
Deal, T.E., 197
Dearmin Huebsch, P., 179
DeKerckhove, D., 31
Delacote, G., 211
De Palma, P., 103
Dery, M., 150
Dickinson, K., 70, 201
Dillman, D.A., 55
DiMaggio, P., 96, 144, 230, 239
Doherty, C., 217
Donath, J.S., 132, 170
Downloading hate, 108
Doyle, T., 52
Dunn, D.G., 153
Duran, R.L., 64, 173
Durso, F.T., 202
Dutton, W.H., 65
Dyrli, O.E., 80

E

Eagly, A.H., 11
Eastin, M.S., 8
Echt, K.V., 135
Education Act, 209
Elkin, T., 79
Ervin, K.S., 136
Ess, C., 8
Estabrook, N., 52

F

Fader, P.S., 201
Farrell, D., 179
Farris, E., 213
Fayol, H., 190, 206
Feldman, A., 216
Feldman, M.S., 201

Feldman, Y., 144
Ferguson, D., 52, 61
Fernback, J., 146, 194, 226
Ferrigno-Stack, J., 143
Festinger, L., 97, 114
Fidler, R., 17, 25, 33, 43, 229
Fiedler, F., 191, 207
Fight hoaxters with the urban legend combat kit, 100
Filkins, D., 91, 101
Fischer, D., 146
Fisher, K., 201
Flanders, V., 52, 149
Flores, F., 120
Floyd, K., 8, 143, 146, 151, 155, 230
Follett, M., 190, 207
Fouser, R.J., 155, 156
Fox, S.A., 133
Fredericks, M., 217
Freed, L., 19
Frey, L., 55
Fulk, J., 8
Funk, J.B., 98, 99

G

Gackenbach, J., 22, 123
Gaffney, G.R., 111, 244
Gale, C., 216
Ganesh, S., 125
Gastil, J., 103
Gaud, W.S., 215, 217
Gavin, J., 156, 158
Geist, P., 191
Geller, L., 229
Gelman, R.B., 99
Gergen, K.J., 138, 139
Gilbert, J., 203
Gilster, P., 15, 52
Goffman, E., 120
Gold, L., 212
Goldhamer, H., 19
Golson, B., 147, 205, 230
Gotthoffer, D., 52
Gozzi, R., 32
Grandgenett, D., 179
Greenhalgh, T., 80
Gregor, P., 105
Grice, H.P., 78
Griffin, E., 151

Grimmes, B., 200
Guernsey, L., 109
Guerrero, L.K., 132
Gulia, M., 230
Gumpertz, J.J., 9
Gurak, L.J., 17, 53, 145, 150, 152, 160

H

Hacker, K., 135, 136
Hackworth, C.A., 202
Hamman, R.B., 8, 230
Hanks, W.F., 171
Hansen, J., 169
Hanson, R., 211
Hargittai, E., 144, 239
Harmon, A., 100, 101, 106, 168, 177
Harnad, S., 121
Harris, J., 52
Harrison, S.A., 244
Hartman, K., 52
Hawisher, G.E., 17
Haythornthwaite, C., 201
Head, S., 214
Heider, F., 29
Heim, M., 15, 37, 151
Henry, P.D., 202
Herbert, B., 109
Herring, S., 8, 126, 152, 182
Hewitt, B., 157
Hill, P., 209
Hiltz, S., 95
Hjelmquist, E., 203
Hobman, E., 24
Hoffman, D.L., 96
Holeton, R., 37
Hollingshead, A.B., 173
Holloway, K., 211
Honeycutt, L., 63
Hoogstra, L., 9
Horton, D., 157
Horton, S., 87
Howard, L.A., 191
Howard, P.E.N., 143
Howe, W., 52
Huizinga, J., 74

I

Infante, D.A., 123
Ingber, D., 130

Introna, L.D., 65
Ivers, K., 52

J
Jackson, C., 179
Jackson, L.A., 96, 136
Jamieson, P., 201
Janis, I., 201
Janney, A.W., 156
Jansma, L.L., 105
Janssen Reinen, I.A.M., 103
Jenkins, H., 134
Johnscher, C., 195
Johnson, K., 109
Johnson, M., 103
Joinson, A.N., 132
Jones, R., 215
Jones, S.G., 2, 8, 15, 28, 53, 149, 150, 171
Jubeck, T., 211
Jung, J., 96

K
Kahn, R.L., 192
Kakutani, M., 101, 110
Kalmbach, J., 216
Kaminer, W., 107
Kanfer, A., 8, 230
Kasket, E., 201
Katz, D., 192
Katz, J.E., 139, 230
Kearney, P., 169
Keil, M., 202
Kelley, M., 137
Kelly, K., 239
Kelly, L., 64, 173
Kendall, N.M., 75
Kennedy, A.A., 197
Kent, M.L., 203
Kent, R.J., 200
Kernal, H.K., 241
Kestnbaum, M., 144, 161, 239
Khan, B.H., 217
Kiesler, S., 10, 24, 151, 153, 158, 201
Kilian, C., 61, 211
Kimmel, M., 9, 25, 111
Klaw, E., 179
Kohut, A., 144
Konold, C., 216
Kopper, G., 52, 61
Kornblum, J., 75

Kramarski, B., 144
Kraut, R.E., 8, 10, 24, 151, 153, 173
Kreuter, M., 179
Kubey, R.W., 88
Kuehn, S.A., 88
Kunde, D., 214
Kushner, D., 80
Kwok-Chi, N., 217

L
Lake, R., 209
Lakoff, G., 103
Lally, V., 137
Lambert, B., 156, 157
Lambert, S., 52
Lamude, K.G., 203
Lane, D.M., 159, 214
Langer, S.K., 123
Lanham, R., 32
LaRose, R., 8
Larsen, R., 203
Lave, J., 171
Lawrence, P.R., 191, 207
Lea, M., 146, 150, 151
Lee, J.Y., 32
Lewin, A., 198, 207
Lewis, L., 213
Lieb, T., 216
Likert, R., 191, 208
Lindsay, R., 215
Lissitz, R.W., 179
Littlejohn, S.W., 123, 132
Lorsch, J.W., 191, 207
Love, G., 8, 151, 162
Lucas, S.E., 77
Lynch, P.D., 87, 200

M
Mabry, M., 106
MacKay, D.G., 103
Maitland, C., 212
Malinek, V., 201
Malone, M.S., 199
Mandel, T., 8
Man posts bail in cyberspace rape case, 107
Manusov, V., 29
Marriott, M., 235
Martin, G., 121
Mayer, D., 217
McCandlish, S., 99

McCollum, K., 214
McCombs, M., 153
McConnell, E., 135
McCormick, J.W., 173, 221
McCormick, N.B., 173, 221
McCown, J., 146
McDermott, I.E., 73
McDonald, M., 109
McGregor, D., 190, 208
McHugh, M.P., 158
McIntosh, D., 76, 242
McKelvey, T., 107
McKenna, K.Y.A., 160, 172, 173, 239
McLuhan, M., 8, 31, 164
McMillan, S.J., 179
McMurdo, G., 24
McRae, K., 132
Medintz, S., 80
Merisotis, J., 91, 92, 211
Messmer, M., 203
Miller, P.J., 9, 215
Miller, S., 217
Miner, A.G., 171
Mitchell, K.J., 143
Moe, W.W., 201
Mok, C., 89
Montgomery, B., 37, 38
Morin, G., 246
Morkes, J., 241
Morrell, R.W., 135
Mouton, J.S., 197
Mowlana, H., 8, 34, 43
Muchmore, M.W., 12
Mukhopadhyay, T., 8
Munger, J., 111
Murray, D.E., 205

N
Nass, C., 75
Neer, M., 123
Negroponte, N., 153
Neher, W.W., 188, 191, 192, 194, 195, 196, 198
Nelson, L.J., 138
Nesson, C., 10
Netchaeva, I., 96
Newhagen, J., 147
Newton, P., 103
Nightline, 149
Nissenbaum, H., 65
Nofsinger, R.E., 128

Nomura Research Institute, 155
Noonan, R.J., 111
Novak, T.P., 96
Nussbaum, J., 135

O
Oder, N., 107
Olsen, F., 107
O'Neil, J., 147
Ong, W.J., 9, 32, 38
Oravec, J.A., 194
Ortner, S.B., 9
O'Shea, T., 150, 151
Ostrom, T.M., 97, 114
Ouchi, W., 190, 208
Ouillette, T., 147

P
Pacanowsky, M.E., 189, 207
Pack, T., 143
Page, L., 143
Paglia, C., 36
Palmer, M.T., 143
Panero, J.C., 159
Papert, S.A., 96, 153
Parker, L., 108
Parks, J.B., 103
Parks, M.R., 8, 143, 146, 151, 155, 230
Pascual, A.M., 199, 205
Paternite, C.E., 180
Patterson, M., 8, 151, 173
Pavlik, J., 24
Pawlowski, D.R., 38, 39
Pecchioni, L., 135
Peiris, D.R., 105
Penn, G., 209
Perlstein, L., 214
Perrewe, P.L., 138
Perse, E.M., 52, 61, 153, 173
Pesce, M., 18, 70, 71
Petersen, A., 52
Peterson, J., 10
Peterson, M.M., 107
Petty, R.E., 97, 114, 131
Phipps, R., 91, 92, 211
Plog, F., 219
Plomp, T.J., 103
Poisoning the Web: Hatred online, 108
Poole, T., 169
Porter, P., 117

Postman, N., 36
Postmes, T., 95
Power, R., 44
The power of speech, 80
Putnam, L.L., 189, 207

Q
Qiu, J.L., 96

R
Radford, M., 52, 53
Rafaeli, S., 63, 64
Rainie, L., 143
Rancer, A.S., 123
Rawlins, W., 38
Ray, D.S., 54
Redding, W.C., 188
Reeves, B., 75
Reid, E.M., 24, 162
Reid, F.J., 201
Rheingold, H., 171, 172
Rice, R.E., 8, 151, 162
Rifkin, J., 150
Riphagen, J., 8, 230
Roberton, M.A., 103
Robinson, J.P., 143, 144, 161, 239
Rocks, D., 199, 205
Rogers, E.M., 41, 45, 231
Rogers, M., 107
Romine, J., 25
Rose, L., 246
Rosenfeld, L.B., 27, 76, 242
Rovner, J., 107
Rubin, R.B., 158, 173
Russell, T.L., 212
Rutkowski, A., 220

S
Saffo, P., 19
Safire, W., 103, 105
Sager, I., 44
Sammons, M., 61
Samoriski, J., 37, 246
Sankis, L.M., 105
Sapir, E., 80
Sarbaugh-Thompson, M., 201
Savage, G.T., 132, 168, 183
Savicki, V., 137
Sayers, J., 11

Schaefer, D.R., 55
Schatz, B.R., 171
Schmidt, S., 214
Schmitz, J., 8
Schneider, S.K., 96
Schofield, J.W., 96
Schonsheck, J., 147
Schroeder, R., 74, 79, 98
Schwartz, J., 112
Science and technology: Picture perfect?, 238
Scodari, C., 75, 76, 97, 242
Sekaquaptewa, D., 103
Selfe, C.L., 17
Shah, R., 25
Shannon, C., 28
Shedletsky, L.J., 24, 120, 121, 122, 124, 125, 128, 130, 147, 173, 213
Shelley, M., 214
Shelton, M., 214
Sheppard, R., 107
Sherer, K., 246
Sherry, L., 216
Shields, S.F., 145, 152
Shockley-Zalabak, P., 23, 187, 194, 232
Sieber, U., 108
Siegel, D., 89
Simmons, J., 203
Skinner, B.F., 203
Skinner, D., 97
Slater, J., 157
Slouka, M., 151
Smith, J.J., 202
Snow, K., 213
Soat, J., 157
Soukup, C., 182
Spears, R., 95, 146, 151
Spinello, R., 246
Sportolari, L., 151
Sproull, L., 151, 158, 201
Staudenmeier Jr., J.J., 153
Stead, B.A., 203
Stefik, M., 80
Steiner, R., 135, 136
Stephan, K., 200
Stephenson, W., 73
Stewart, C.M., 145, 152
Stix, G., 201
Strand, E.A., 103
Stross, R.E., 100, 110

Sudweeks, F., 63, 64
Suler, J., 151, 152, 205
Sveningson, M., 158
Swats, V.R., 143
Sypher, B.D., 188

T

Taguiri, R., 197, 198, 207
Tannen, D., 38
Tanner, E., 64
Tao, L., 128, 218
Taylor, D., 151
Taylor, F.W., 189, 208
Teich, A.H., 136
Teske, J., 201
Tewksbury, D., 182
Thatcher, J.B., 138
Thomas, E., 106
Thomas, J., 109
Thompsen, P.A., 105
Thoreau, H.D., 41
Thorington, H., 180
Townsend, A.M., 204
Turkle, S., 37, 71, 74, 76, 128, 140, 146, 151, 157, 158
Turoff, M., 95

U

Ullman, E., 7, 150
Uslaner, E., 239

V

Valanides, N., 201
Van Der Leun, G., 8
Van Dijk, J., 136
Veryard, R., 80
Vitanza, V., 7
VonBertalanffy, L., 192
von Hippel, W., 103
Vygotsky, L.S., 103, 126

W

Wajcman, J., 96, 136
Wakabayashi, R., 91
Waldeck, J., 169
Walsh, N.E., 19
Walston, J.T., 179
Walters, T.L., 143

Walther, J.B., 8, 33, 143, 144, 151, 161
Waterman, R.H., 199
Watt, D., 122, 153
Weaver, W., 28
Weber, M., 189, 207
Wegner, S., 211
Weick, K., 207
Weiner, R., 217
Weisband, S.P., 96
Welch, K.E., 21
Wellman, B., 201, 230
Wellner, A.S., 96
Wenger, E., 171
Werde, B., 157
Wheeler, D., 144
White, H., 135
White, J.M., 122, 153
Whitelaw, K., 107
Whitley, B.E., 103
Whittle, D.B., 162
Whitty, M., 156, 158
Whorf, B.L., 80
Widiger, T.A., 105
Wiener, N., 37
Wildermuth, S., 157, 158
Wildstrom, S.H., 235
Wilhelm, A.B., 7
Willis, M., 52, 149
Winch, P., 124
Winograd, T., 120
Winston, B., 17, 33, 41
Winterowd, C.L., 154
Wohl, R.R., 157
Wolak, J., 143
Wood, J.T., 123
Woolley, B., 36
Wright, K.B., 178
Wynn, E., 139

Y

Year 2000, 214, 217
Young, K.S., 101, 102

Z

Zagorsky, J.L., 217
Zillmann, D., 10

Subject Index

A

Academic collaboration, 217–219
Administration, and theories about the workplace, 190
Adoption, 41–42
Advanced Research Projects Agency (ARPA), 246
Age differences in intrapersonal communication, 134–136
Agenda Setting Theory, 153
Amazon.com, 199
Ambiguity, 139
America Online (AOL), 108, 160
Anonymity, 175, 182
Archives, 67
Argot, 183
Association, 32
Asymmetric information, 67
Attributing human characteristics to computers, 130–131
Attributing meaning, 29, 32

B

Behavioralist, 45
Blackboard, 6, 13, 160
Boolean search, 59, 67
Boundaries, 206
Broadcasting, 45
Bureaucracy, 189
Business communication on the Internet, 191–192

C

California Department of Education, 209
CD/DVD burner, 246
Channels, 28, 32
Chat rooms, 169, 183
Classical management, 190, 207
Climate, 197–198, 207
Cognitive collaboration, intrapersonal communication as, 117–140
Cognitive dissonance, 97–98
Collaborative communication in online business interactions, 201
Communicating scholarly research online, 62–63
Communication, 27, 34–46, 86–87, 124, 131–132, 191–192
   consumer, in online business interaction, 200–201
   design on the web, 86–87
   flow, 207
   intensification, table, 27
   interactive, 131–132
   on the Internet, 15–33, 166–168, 191–192
      business, 191–192
      human, 15–33, 166–168
      tensions of, 34–46
   interpersonal, 141–163
   intrapersonal, 117–140
   managerial, 203
   media, 25–27
      language and writing, 25
      visually oriented, 26–27
   organizational, 188–189
   overload, 202–203, 207
   patterns in the workplace, 193
   perspective, 229–233
   quality, primary sources about, 55
   research, 52–53
   scholarly, 53–54
Communication and Aging, 135–136
Communication Teacher Resources Online (NCA), 63
Community, 143, 163
Computer games, 79
Computer-mediated communication (CMC), 14, 20–21, 32, 95, 143, 192–193, 195, 214
   paradoxical nature of, 14

Computers, 152–153, 234–236
  in the future, 234–236
  influence of, on family, 152–153
Connected Intelligence: The Arrival of the Web Society (DeKerckhove), 31
Consciousness, 13, 32
Consequences and conclusions, 225–246
Consumer communication in online business interaction, 200–201
Context, 22, 32, 124, 139
Contingency theories, 191, 207
Contradiction, 11
Convergence, 13, 228–229, 246
Conversing, 42–43
Course Compass, 6, 13, 62
Course environment, 13
Critical perspective, 189, 207
Critical reading, 54
Cultural homogenization, 8, 13
Cultural mediational artifacts, 124
Culture, 197–198
  differences in intrapersonal communication, 133–134
Cyberliteracy, 21–23, 32
Cyberspace, 36–37, 45
Cyberspace playground, 246

D
Databases, online, 55–57, 67
Decision making, 195–196
Decoding, 28, 32
Democratization, 114
Developing community through online groups, 171–172
Diffusion, 41–42, 45
Digital divide in families, the Internet and, 95–97
Discourse, 32
Discourse analysis, 45
Discussion groups, 94, 114
  online scholarly, 63–64
Distance education, 211, 222
Dot com, 207

E
E-business, 199–200, 207
E-discussion, 67
Education Act, 209
Educational contexts, 209–222
  academic collaboration, 217–219

educational policy and Internet communication, 217
intensification of effort, 215–217
intensified learning, 211–212
online *versus* traditional classroom education, 212–215
Effects, 32
E-journals, 60–61, 67
Electronic discussion, 169, 183
Electronic Journal Miner, 60
Electronic workflow, 199, 207
E-mail, 146–147, 163, 174, 177–178
  high volume, 175
  ignoring, 174
Empirical research, 13
Empowering self, 167–168
Encoding, 28, 32
End users, 14
Enhancing relationships with family and friends online, 146
ERIC (Educational Resources Information Center), 222
Ethnomethodology, 45
Expanded Academic Index, 222

F
Face-to-face (f2f or FtF) communication, 143, 163
Facing fears, Internet as a tool for, 110
Family, 152–155
  Internet as a source of communication for, 153
  new opportunities for interpersonal relationships, 154–155
  reinforcing a child's self-esteem, 154
  using Internet to increase human interplay, 153–154
Fantasy themes, 89
Fear of terrorism, 109–110
Feedback, 28, 33, 37, 45
First Look at Communication Theory, A (Griffin), 62
Flaming, 105–106, 114, 150, 163
Flat organization, 207
Full-text article, 68
Full-text databases, 57
Functional perspective, 188–189, 207

G
Gender differences in intrapersonal communication, 136–137

Globalization, 36, 45
Global village, 14
Google.com, 143, 172
Groups, 164–183
   characteristics of online group discussion, 172–177
      conversation lulls, 174–175
      high volume e-mail, 175
      ignoring, 174
      Internet anonymity, 175
      lack of nonverbals, 176–177
      lurking, 174
      unique language, 173
      virtual reality, 176
   considerations for moderating online discussions, 177–178
      allowing confidential talk, 178
      avoiding private, direct e-mails, 178
      creating a pragmatic system in advance, 177
      encouraging brief e-mails, 177
   developing community through, 171–172
   health care empowerment through, 178–180
      accuracy of information, 179–180
      patient-doctor communication, 179
      patient storytelling groups, 179
   mediator of human communication, 166–168
      connecting people with needs or resources, 166–167
      empowering the self, 167–168
      improving the quality of life, 168
   storytelling in, 168–171
Groupthink, 140, 202, 207

H

Hate sites, 108
Hate speech, 107–109, 114
Health care empowerment through online groups, 178–180
   accuracy of information, 179–180
   patient-doctor communication, 179
   patient storytelling groups, 179
Hierarchy, 207
Hit, 68
Hoaxes, 100–101
Hostile metaphors, 102–105
Human communication on the Internet, process of, 15–33, 166–168
   communication media, 25–27
      language and writing, 25
      visual orientation, 26–27
   developmental perspective, 17–23
      cyberliteracy, 21–23
   Internet as mediator, 166–168
      connecting people who have needed info or resources, 166–167
      empowering self, 167–168
      improving quality of life, 168
   models of communication, 28–29
   reasons for studying, 23–24
Human-computer interaction (HCI), 20, 33, 75
Humanist, 45
   versus behavioralist approach, 38–41
Human relations, 190

I

Identity role-playing, 75–76
Imaginative play like a child, 75
Immediacy, 143, 163
Informatics, 49–69
   communicating scholarly research online, 62–63
   defining, 51–52
   e-journals, 60–61
   journalism and Internet scholarship, 61–62
   online databases, 55–57
      full-text databases, 57
   online scholarly discussion groups, 63–64
   online search, 57–60
   paradoxes of, table, 64
   quality, primary sources about communication, 55
   research communication, 52–53
   scholarly communication, 53–54
      critical reading, 54
      peer reviewed journals, 54
      refereed journals, 54
Information technology (IT), 68
Informing, 42–43
Ingroup, 170
Inner-outer dichotomy, 45
Inner speech, 122–123
Instant messaging (IM), 124–126, 169, 183
Intensification, 33
   of polarization, 94–95
Intensified interplay, 70
Intensified learning, 211–212
Intention, 140
Interaction involvement, 168, 183
Interactionism, 140
Interactive communication, 131–132

Interactivity, 68
Internet, 20, 22, 33, 84–85, 101–102, 234–236
  addiction, 101–102, 114
  anonymity, 175
  communication, 217, 234–236
    educational policy and, 217
    in the future, 234–236
  and communication principles, interplay between, 84–85
  and digital divide in families, 95–97
  flaming, 105–106
  games for destruction, 76–77
  hoaxes, 100–101, 114
  instances of use, 22
  interpersonal communication, 141–163
  intrapersonal communication, 119–123
  metaphors, table, 104–105
  play interfering with business, 196
  relationships, stages of, figure, 159
  role-playing on the, 132–133
  self-identity and the, 127–130
  tensions of communication on the, 34–46
  as a tool for facing fears, 101–102
  video games, 98–99
  and violence, 106–109
Interpersonal communication, 9, 14, 141–163
  influence on family, 152–155
    Internet as a source of communication, 153
    new opportunities for interpersonal relationships, 154–155
    reinforcing a child's self-esteem, 154
    using Internet to increase human interplay, 153–154
  and Internet communication,
    relationship between, 143–148
    enhancing relationships with family and friends, 146
    maintaining long-distance relationships, 146–148
    meeting people and creating relationships, 146
  in online business interaction, 201
  online relationships, 155–158
  speed, reach, anonymity,
    regard, and interactivity, 149–152
Interplay between Internet and communication principles, 84–85
Interpretive perspective, 189, 207
Intervening variables, 20, 33

Intrapersonal communication, 9, 14, 117–140
  age differences, 134–136
  attributing human characteristics to computers, 130–131
  culture differences, 133–134
  gender differences, 136–137
  interactive communication, 131–132
  and the Internet, 119–123
    inner/outer speech, 122–123
  resolving the inner/outer dichotomy, 124–127
  role-playing, 132–133
  self-identity and the Internet, 127–130

J
Journalism and Internet scholarship, online research, 61–62

L
Language of discussion groups, 173
Linguistics, 45
Listserv, 169
Lurking, 174, 183

M
Maine Department of Education, 214
Maintaining long-distance relationships online, 146–148
Management information systems (MIS), 18
Managerial communication in the workplace, 203
Mass communication, 10, 14
Meaning, 45
Meaning-making, 14
Mediated symbolic process, 123
Mediation, 124
Mediator of human communication, Internet as, 166–168
  connecting people with needs or resources, 166–167
  empowering the self, 167–168
  improving the quality of life, 168
Medium, 23, 33
Meeting people and creating relationships online, 146
Messages, 28, 33
Metacognition, 121
Metamessage, 128
Metamorphosis, 17, 33

Metaphors, 90, 114
Methodology, 45
Mind, 124, 140
Models of communication, 28–29
Moderating online discussions, considerations for, 177–178
   allowing confidential talk, 178
   avoiding private, direct e-mails, 178
   creating a pragmatic system in advance, 177
   encouraging brief e-mails, 177
MOOs, 75–76, 90
MUDs, 74–75, 90

N

Narrowcasting, 46
National Communication Association (NCA), 63
National Research and Education Network (NREN), 53
National Science Foundation (NSF), 246
Netspeak, 102, 114
Nightline, 149
Noise, 28, 33
Nomura Research Institute, 155

O

Observable variables, 46
Onground, 46
Online, 55–57, 68, 200–203, 211–215, 222
   business interaction, 200–203
      collaborative communication, 201
      communication overload, 202–203
      consumer communication, 200–201
      groupthink, 202
      interpersonal communication, 201
      risky communication, 203
   databases, 55–57
   education, 211–215, 222
      versus traditional classroom education, 212–215
   enhancing relationships with family and friends, 146
   group discussions, 172–178
      characteristics of, 172–177
      considerations for moderating, 177–178
   groups, developing community through, 171–172
   interpersonal communication, paradoxes of, table, 160
   relationships, 146–148, 155–158
      maintaining long-distance, 146–148
      meeting people and creating, 146
      stages of, figure, 159
   scholarly discussion groups, 63–64
   search, 57–60
Operationalization, 46
Organizational communication, perspectives of understanding, 188–189
Organizational culture/climate, 197–198
Outcomes-based education, 222
Outer speech, 122–123

P

Partially distributed work groups, 208
Participative decision making, 191, 208
PDF files, 68
Peer reviewed journals, 54
People and organizations, symbiotic relationship between, 191
Personal digital assistant (PDA), 246
Plagiarism, 68
Play, 70–90
   of the Internet, 72–78
      computer games, 79
      the fun of something new, 77–78
      games for destruction, 76–77
      identity role-playing, 75–76
      imaginative play like a child, 75
      interplay with self, 76
      playful work, 77
      play like theater, 76
   of Internet communication, 70–90
      communication design on the web, 86–87
      example metaphor, 81–84
      interplay between Internet and communication principles, 84–85
      playful metaphors, 79–81
Play Theory of Mass Communication (Stephenson), 18
Polarization of people, 91–114
   cognitive dissonance, 97–98
   digital divide, 95–97
   facing fears, 110
   fear of terrorism, 109–110
   flaming, 105–106
   hate speech, 107–109
   hoaxes, rumors, and myths, 100–101
   hostile metaphors, 102–105
   intensification of polarization, 94–95
   Internet addiction, 101–102

Polarization of people, (continued)
   personal fear, 106–107
   privacy, 99–100
   video games, 98–99
Portals, 114
Postcyberdisclosure panic (PCDP), 162
Preservice teacher education, 222
Primary sources of information, 55, 68
Privacy, 10, 99–100
Process of Communication, The (Berlo), 18
Professionalism, 208
Public Speaking Unbound, 62

Q
QuackWatch.Org, 180
Quality of life, improving, 168
Quality sources of information, 55, 68

R
Receiver, 28, 33
Reconfiguration, 163
Refereed journals, 54, 68
Regard, 163
Relational communication, 163
Reliability, 68
Research, 68
   communication, 52–53
   Risky communication in online business interactions, 203
   Role-playing on the Internet, 132–133
   Roles, 208
   Rumors, 100–101

S
Scanning, 68
Scholarly, 53–54, 68
   communication, 53–54
   discussion groups, 68
   periodicals, 68
   research, 68
Scientific management, 189–190, 208
Search engine, 68
Second self, 140
Second source, 69
Smart phones, 168, 246
Storytelling in groups, 168–171
Student as consumer, 222
Subjectivity, 46
Supportive education, 211, 222

Survey technique, 69
Symbolic relationship, 208
Symbolism, 73, 90
Symbols, 46
Synchronous interaction, 90
Syntax, 33
Systems, 19, 33
   theory, 208
   workplace, 192

T
Technological determinism, 17, 33
Telecommunications Act (1996), 19
Tensions, 34–46, 195–196
   of communication on the Internet, 34–46
     adoption and diffusion, 41–42
     cyberspace, 36–37
     globalization, 36
     humanist *versus* behavioralist approach, 38–41
     informing and conversing, 42–43
     theoretical roots of, 37–38
   workplace, 195–196
     decision making, 195–196
     Internet play interfering with business, 196
     paradoxes regarding the Internet and organizations, 195
Terrorism, 109–110, 114
Textbooks, online sites, 62
Theory, 46
Theory X and Y, 190, 208
Thirty-year rule, 19, 32
Tone, 140
Topics, Inc., 172
Trade magazines, 69
Transformational grammar, 46
Turn-taking, 140

U
Understanding Media: The Extensions of Man (McLuhan), 18, 31
Unique language of discussion groups, 173
Urban legends, 100, 114

V
Validity, 69
Video games, 98–99
Virtual Culture: Identity and Communication in CyberSociety, 152

Virtual reality, 176, 183
Voice, 140

W
Webboard, 169
WebCT, 14, 129, 160
Web page, 90
Workplace, 187–208
   business communication on the internet, 191–192
   communication patterns, 193
   culture and climate, 197–198
   e-business, 199–200
   manager communication, 203
   online business interaction, 200–203
      collaborative communication, 201
      communication overload, 202–203
   consumer communication, 200–201
      groupthink, 202
      interpersonal communication, 201
      risky communication, 203
   organizational communication, perspectives of understanding, 188–189
   revising traditional theories, 189–191
      administration, 190
      bureaucracy, 189
      classical management, 190
      contingency theories, 191
      human relations, 190
      participative decision making, 191
      people and organizations, symbiotic relationship between, 191
      scientific management, 189–190
      Theory X and Theory Y, 190
   roles, 193–195
   systems, 192
   tensions, 195–196
      decision making, 195–196
      Internet play interfering with business, 196
      paradoxes regarding the Internet and organizations, 195
World Wide Web (WWW), 20, 33

Y
Yahoo.com, 172